Michael Decker, Mathias Gutmann (Eds.)

Robo- and Informationethics

Hermeneutics and Anthropology
Hermeneutik und Anthropologie

edited by/herausgegeben von

Prof. Dr. Andrea Marlen Esser
(Universität Marburg)

Prof. Dr. Armin Grunwald
(Karlsruhe Institute of Technology – KIT)

Prof. Dr. Dr. Mathias Gutmann
(Karlsruhe Institute of Technology – KIT)

Volume/Band 3

LIT

Robo- and Informationethics

Some Fundamentals

edited by

Michael Decker and Mathias Gutmann

LIT

This work has been funded by the New Field Group "Autonomous Systems" and the KIT Focus "Humans and Technology", which received financial support from the "Concept for the Future of Karlsruhe Institute of Technology (KIT)" within the framework of the nationwide "Initiative for Excellence".

Gedruckt auf alterungsbeständigem Werkdruckpapier entsprechend
ANSI Z3948   DIN ISO 9706

Bibliographic information published by the Deutsche Nationalbibliothek
The Deutsche Nationalbibliothek lists this publication in the Deutsche Nationalbibliografie; detailed bibliographic data are available in the Internet at http://dnb.d-nb.de.

ISBN 978-3-643-90208-5

A catalogue record for this book is available from the British Library

©LIT VERLAG GmbH & Co. KG Wien,   LIT VERLAG Dr. W. Hopf
Zweigniederlassung Zürich 2012   Berlin 2012
Klosbachstr. 107   Fresnostr. 2
CH-8032 Zürich   D-48159 Münster
Tel. +41 (0) 44-251 75 05   Tel. +49 (0) 2 51-620 320
Fax +41 (0) 44-251 75 06   Fax +49 (0) 2 51-23 19 72
e-Mail: zuerich@lit-verlag.ch   e-Mail: lit@lit-verlag.de
http://www.lit-verlag.ch   http://www.lit-verlag.de

Distribution:
In Germany: LIT Verlag Fresnostr. 2, D-48159 Münster
Tel. +49 (0) 2 51-620 32 22, Fax +49 (0) 2 51-922 60 99, e-mail: vertrieb@lit-verlag.de
In Austria: Medienlogistik Pichler-ÖBZ, e-mail: mlo@medien-logistik.at
In Switzerland: B + M Buch- und Medienvertrieb, e-mail: order@buch-medien.ch
In the UK: Global Book Marketing, e-mail: mo@centralbooks.com

Contents

Robo- and Informationethics: Some Introducing Remarks 3
Michael Decker & Mathias Gutmann

Armed Robots and Preventive Arms Control 7
Jürgen Altmann

Investigating the Robot in the Loop. Technology Assessment
in the Interdisciplinary Research Field Service Rotics 31
Martin Meister

Technology Assessment of Service Robotics.
Preliminary Thoughts Guided by Case Studies 53
Michael Decker

Ethical Aspects of Autonomous Systems 89
Herman T. Tavani

Anthropocentric-Based Robotic and Autonomous Systems:
Assessment for New Organisational Options 123
António Brandão Moniz

Ethical and Critical Views on Studies on Robots and Roboethics ...159
Makoto Nakada

Can Robots Plan, and What Does the Answer to this Question
Mean? 189
Armin Grunwald

Between Innovative Forms of Technology and Human Autonomy:
Possibilities and Limitations of the Technical Substitution
of Human Work 211
Peter Janich

Action and Autonomy:
A Hidden Dilemma in Artificial Autonomous Systems 231
Mathias Gutmann, Benjamin Rathgeber & Tareq Syed

Robo- and Informationethics: Some Introducing Remarks

Michael Decker & Mathias Gutmann

Our ideas and concepts of human beings and their environment are deeply intertwined with technical knowledge. This holds true for the concept of the bio-cultural being *Homo sapiens*, at least because it can be described as a descendant of tool-making predecessors (for example *Homo habilis* and *Homo ergaster*, their scientific names already indicating characteristic abilities of these early humans). But despite the characterisation of humans as tool-makers, it is not the mere fact of tool-making itself that defines humans. This fact is obviously useful in order to discern types of material cultures; a well-known procedure even for human predecessors in palaeoanthropology and archaeology (as the terms of Acheuléen, Moustérien or Gravettien may paradigmatically show). It rather seems to be the *wise* of tool-making, the *form* of handling artefacts and the *structure* of the life-forms, which constitute the connection between humans and their tools and means:

> "Man's outstanding characteristic, his distinguishing mark, is not his metaphysical or physical nature – but his work. It is this work, it is the system of human activities, which defines and determines the circle of "humanity." Language, myth, religion, art, science, history are the constituents, the various sectors of this circle. A "philosophy of man" would therefore be a philosophy, which would give us insight into the fundamental structure of each of these human activities, and which at the same time would enable us to understand them as an organic whole." (Cassirer 1972: 68)[1]

Understanding technique as the form of human action, technologies became the actual tools, which are undoubtedly produced by men; but at the same time, these tools alter and change human environment as well as the humans themselves. "Culture" then is not just transformed "nature", depending on human purposes. In descriptions of this kind, humans, technology and culture would have been thought of as emerging *within* nature, as a natural extension of nature itself. This

---

[1] Cassirer, E. (1972): *An Essay on Man*, Yale University Press, New Haven, p. 68.

reduced idea of human being and their world reflected the resulting relation as a "natural" relation, humans fundamentally as a natural unit, founded especially by concepts of life sciences.

The direct opposite – and to a certain extent even the mirror image – to this nature-centred concept of human beings would then be the assumption that humans literally are what they are transforming themselves into. Nature then became a resource of human self-transformation and the difference between nature and culture then would not be a sortal but an aspectual difference, referring to the very form of human action.[2] However, being juxtaposed in premise and consequence, both concepts of humans are closely related and mutually generate each other. If both relations, the nature-culture as well as the culture-nature relation, are separated from each other and the connecting reference – namely the system of human activities – is lost, the identity of humans inevitably becomes doubtful. According to the increasing process of technical self-substitution, a process, which is fuelled and accelerated by a variety of "converging technologies", it is not only the concept of humanity which is threatened immanently, but the concept of nature itself. Thus, converging technologies promote a *correlated* double process, the results of which can be particularly observed in the field of bio- and information technologies. They build an intriguing complex of capabilities and potentialities, which serve as a resource for the replacement and support of human features and activities. In consequence, human capabilities and human skills which root in the "natural" equipment of *Homo sapiens*, become the target of a transformation that finally tends to *transcend* the limits of the original "natural" constitution. This process of self-objectivation and -transformation endangers the relevance of corporeal as well as mental integrity, which both are assumed to be central criteria of the personal existence of *Homo sapiens*. As a result of this self-transformation of human beings "in the age of their artificial reproduction", *Homo sapiens* itself more and more becomes a part and sometimes even an aspect of the environment; an environment which is – in the sense of ambient technologies – defined by the interfaces of information systems that are the less visible and

---

[2] Gutmann, M. (2002): "Human Cultures' Natures", in: Grunwald, A., Gutmann, M. & Neumann-Held E.-M. (eds.): *On Human Nature. Antropology: Biological and Philosophical Foundation*, Springer, Berlin, Heidelberg, New York, pp. 190–240.

observable the more they become functionally optimised. However, the new technologies do not only provide severe anthropological problems, they are of methodological as well as ethical interest, too. The methodological dimension starts with the seemingly simple question whether or not "converging" technologies are of an entirely different nature compared to "traditional" technologies such as internet-based actor-sensor nets, Bayesian nets, gene technology and bioinformatics. The determination of the status and the relevance of these types of technologies, the validity, which can be claimed for the resulting descriptions and the adequacy assumed for the resulting models are then the main purpose of methodological reconstruction. Reconstructions of this type will finally allow us to decide whether the attributes used in terms of e.g. autonomous systems, artificial moral agents or evolutionary robotics indicate proper characterisations of the respective devices (what they undoubtedly are) or mere metaphoric expressions. The clarification of the anthropological and the methodological aspects of those expressions are necessary prerequisites for the proper treatment of normative questions. These kinds of questions are immediately connected with the status of the technological devices (e.g. as "artificial persons") and the consequences and effects that the application of these technologies actually or potentially has or might have on societies which are exposed to severe challenges of different types at the same time. By paradigmatically considering the on-going debate on aging populations and the impact of the aging process on health care as well as warfare, the – possible – role of robotics becomes an important normative and ethical issue. As we are facing an extremely versatile field of research that undergoes the most rapid transformations, it is only reasonably to emphasise that the results of the anthropological, methodological and ethical reconstructions will of course differ fundamentally, depending on the perspective of the reconstructing scientist as well as the specificities of the techniques scrutinised. From this perspective, robo- and informationethics seem to provide exactly the type of reasoning which is necessary in order to reevaluate and determine some constituents of human self-understanding. However, this type of reasoning has to integrate normative as well as methodological aspects and thus this book presents the outcomes of a symposium on the limits of human self-substitution based upon some paradigmatic examples of the development of hybrid technology.

This symposium was organised at KIT with funding received form the KIT Focus "Humans and Technology". The results of the symposium are the very building blocks of the present publication. The first section presents papers, which deal specifically with robotic applications that go far beyond the robots successfully used in the field of industrial production. The second section focuses on normative aspects of roboethics, critically evaluating the description of autonomously acting technical machines as well as normative principles, which are assumed to be relevant for the ethical reflexion of robotic technology. The third section finally presents some methodological reconstruction specifically asking for the validity, which can be claimed for descriptions of robotic agents in the role of human actors.

Michael Decker   Mathias Gutmann

Armed Robots and Preventive Arms Control

Jürgen Altmann[1]

Abstract: Uninhabited vehicles (UVs), used by armed forces since a long time mainly for surveillance, are now being equipped with weapons, with the USA in the lead, using them routinely for attacks in the Middle East. Autonomous killing by machine decision is taken into view already. Uncontrolled proliferation of armed UVs can bring dangers for arms control agreements, international humanitarian law, military stability between potential opponents and security within societies. Preventive arms control is advisable; options for international limitations are presented.
Keywords: uninhabited vehicle/unmanned vehicle; arms control; international humanitarian law; proliferation

1. Introduction

Military work for robots, or vehicles without crew on board, has a long history (Newcome 2004, Gage 1995). Already before 1900 the USA experimented with uninhabited aircraft, in Germany the first flight without a pilot took place in 1903. US work continued in World War I. In World War II, Nazi Germany produced the winged bomb V-1, a precursor to the cruise missile, in large numbers, attacking cities in Belgium and England.

When, starting in 1957, satellites were launched in increasing numbers, most of them did not carry humans. Beginning in the 1960s, (land) robotics research at US universities was funded by the military. Even though piloted aircraft dominated in the Vietnam war, the US forces did thousands of surveillance operations by uninhabited aircraft. Such craft got a boost with the deployment in the 1980s of long-range cruise missiles (carrying nuclear or conventional warheads). At about the same time many armed forces started to deploy uninhabited (or unmanned) air vehicles (UAVs) for reconnaissance purposes.

Since then, UAVs have steadily gained importance, with more than 50 countries developing or producing them. Land vehicles are still at the development and testing stages, mainly because navigation and obstacle avoidance on land pose much higher hurdles than in the air. Development of uninhabited vehicles on and under water is underway,

---

[1] Project on armed uninhabited vehicles funded by German Foundation for Peace Research (DSF).

with motor boats already being offered. In particular the US pushes much wider use of uninhabited vehicles by its military, with routine use in its present wars in Afghanistan, Irak and Pakistan, but the same trend can be observed internationally.

2. Definitions Are Important

Many terms have been used to denote uninhabited vehicles (UVs), for example for aircraft, remotely piloted vehicles, drones or aerial robots. Since a few years the US Department of Defense (DoD) is unifying its language, speaking of "unmanned air/ground/maritime (/surface/undersea) vehicles" (UAV/UGV/UMV(/USV/UUV)) and the related systems (UAS etc.), the system notion indicating that beside the vehicles proper there are additional components such as a ground control station, a data link etc.

The Department of Defense has given a definition which intentionally excludes certain types (DoD 2007: 1, emphasis original):

> "**Unmanned Vehicle.** A powered vehicle that does not carry a human operator, can be operated autonomously or remotely, can be expendable or recoverable, and can carry a lethal or nonlethal payload. Ballistic or semi-ballistic vehicles, cruise missiles, artillery projectiles, torpedoes, mines, satellites, and unattended sensors (with no form of propulsion) are not considered unmanned vehicles. Unmanned vehicles are the primary component of **unmanned systems**."

The exclusion of missiles, torpedoes and satellites makes only limited sense – technically, they are vehicles (that is, mobile objects capable of changing their trajectory) not having a human on board. It constrains the notion of unmanned (or uninhabited)[2] vehicles to the types that are newer, but many similarities exist and will likely increase in the future. For example, cruise missiles may get a capability to be recalled or re-programmed during flight while some UAVs take off from a launcher. New uninhabited underwater vehicles (UUVs) can be quite similar to torpedoes which already now can follow a search pattern or can be controlled via a glass fibre.

More systematic is a general definition:

---

[2] This is the gender-neutral name preferred here.

> "*Uninhabited vehicles* are controllable, usually powered vehicles which do not carry a human operator. A *vehicle* is an (at least partly) artificial object which can manoeuvre in some way."

Based on this wide understanding one can then introduce subcategories concerning criteria such as the ability to loiter, the generality of the target set, the degree of change possible after launch, and of course the medium – air, ground, water, outer space. Such subcategories are also required in defining limitations and verification methods.

For armed uninhabited vehicles, this understanding excludes unguided artillery grenades and bullets, tethered aerostats[3] and trained animals as well as immobile armed robots[4]. Included are aerodynamic and ballistic missiles as well as torpedoes, satellites with orbit-correction engines and electronically controlled insects.

Because change will ensue from the UV categories that are newly introduced, it makes sense to focus on them.

3. Military Non-Weapon Uses of Uninhabited Vehicles

Up to now, the military is the most important user of uninhabited vehicles (UVs) by far.[5] For the time being this means mostly air vehicles which are mainly used for surveillance, with cameras for visible or infrared light or radars. The number of types and the numbers deployed are steadily increasing – more than 50 countries develop or produce UAVs, hundreds of types exist (Jane's 2007, van Blyenburgh 2007). The sizes vary from large (Global Hawk, USA, 40 m wingspan) via medium (HERTI, UK, 12 m wingspan) to small (ALADIN, Germany, 1.5 m wingspan, hand-launched). Prototypes of micro air vehicles with wingspans down to 10 cm have flown already, some relying on flapping wings instead of propellers.

---

[3] One could argue that a tethered aerostat can do limited manoeuvres by changing the length of its tether or moving its mooring station. Such "vehicles" are included in the Unmanned Systems Roadmap (DoD 2009: 87ff.).

[4] Immobile armed robots could produce special, though limited, problems and would have to be covered by specific regulation.

[5] The only exceptions are in underwater systems where remotely operated vehicles are in routine use for oil and gas platforms and underwater pipelines, and in satellites where a significant share is used for civilian communication, weather observation, broadcasting etc.

In uninhabited ground vehicles (UGVs), recent years of US warfighting in the Middle East have seen a surge in small (0.5-1 m) robots for clearing of roadside bombs, remotely controlled from tens of metres distance – in 2009, more than 6,000 such robots had been deployed (DoD 2009: 3).

In uninhabited surface vehicles (USVs), military motor boats (size 5 to 12 m) have been developed for reconnaissance, submarine detection and mine countermeasures. In uninhabited underwater vehicles (UUVs), work focuses on torpedo-like systems from 1.5 to 3 m length, mainly for detecting mines and submarines.

Outer space is special – first, there is no atmosphere that could create a lift force and satellites have to move very fast to not fall down to Earth. Second, nearly all satellites are uninhabited.

In the next 10-20 years a continously growing world market for various types of UVs is expected, still mostly consisting of military systems. Additional markets are possible for non-military state applications – such as border control, disaster monitoring, police – and for private industry, for example for small UAVs for checking high-voltage power lines. Efforts for allowing UAVs into general air space are underway on many levels. Household and other private-user service robots, also expected to increase strongly, will have limited mobility and less overlap with military UVs.

The US DoD is the most important driver of UV developments worldwide – it is active in forming a joint approach for all services, that is, Army, Navy and Air Force. For this purpose it has published roadmaps spanning 25 years, one in 2007 and a new version in 2009 (DoD 2007, 2009). Part of the efforts planned is equipping UVs with weapons (Singer 2009, Krishnan 2009).

4. Trend Towards Arming of Uninhabited Vehicles

A major step in this direction was the directive by the US Congress in 2000 that in 2010 one third of the operational deep-strike aircraft and in 2015 one third of the operational ground-combat vehicles shall be unmanned (DoD 2007: 6). Driven by the US wars in Iraq, Afghanistan and elsewhere in the Middle East, the Air Force retrofitted its Predator surveillance UAV (8 m length) with two Hellfire laser-guided air-to-ground missiles, first used to kill six putative terrorists in a car in Yemen in 2002 (Hoyle & Koch 2002). Next, a bigger type was developed (Predator B, later called Reaper, 11 m length), carrying up to 1,360 kg of weapons: Hellfire missiles, Sidewinder air-to-air missiles, Paveway II laser-guided bombs and the Joint Direct Attack Munition. By early 2009, the Air Force had 195 Predators and 28 Reapers (Drew 2009). They are routinely used to attack Taliban and other insurgents, under remote control from air bases back in the US. In many cases, non-combatants have been killed – a famous first example occurred in February 2002, when CIA operatives spotted a tall man in Afghanistan and, taking him as Osama bin Laden, killed him with two other men. However, they were local people gathering scrap metal (Hersh 2002).

In recent years, such attacks have markedly increased. In Pakistan alone there were 33 in 2008, 53 in 2009 and 118 in 2010 (New America Foundation 2011, see also Weber 2009). In addition to attacks by the military, the CIA uses Predator and Reaper UAVs for targeted killings, in particular in Pakistan. Philip Alston, the United Nations Special Rapporteur on Extrajudicial, Summary, or Arbitrary Executions, has questioned the compatibility with the laws of warfare (also called international humanitarian law) and human rights law; also here non-combatants become victims often (Mayer 2009, Alston 2010).

These armed UAVs are but a first step. For the future, UAVs are envisaged for all forms of combat and its support that piloted aircraft used to do. For example, the DoD foresees UAVs for aerial refueling in 2024 and for aerial combat in 2032 (DoD 2009: 18). Prototypes of such uninhabited combat air vehicles (UCAVs) are being developed and built not only in the USA (UCAS-D/X-47B), but also in France (nEuron, with Sweden, Greece, Switzerland, Spain, Italy), Germany (Barracuda, with Spain), United Kingdom (Taranis) and Russia (Skat). At the other end of the size scale, micro air vehicles can also be armed – reportedly the British Special Air Service has put explosive on Wasp aircraft (wing span 41 cm) for "kamikaze" attack on Taliban snipers; the USA and Israel are thinking along similar lines with markedly smaller UAVs (Hambling 2007, 2010).

A special concept is the so-called prompt global strike – here a missile is to be driven at several times the speed of sound through the air to hit any place on Earth with conventional munition within an hour; whether it will be realised is open (Sanger/Shanker 2010).

Concerning ground vehicles, the SWORDS small robot and its successor MAARS can be equipped with an assault rifle, machine guns or grenade launchers (Jewell 2004, Qinetiq-Foster-Miller 2010). It was deployed to Iraq in 2008 but not used in combat. On a much larger scale, the US Army had an extremely ambitious programme, the Future Combat System (FCS), since 2000/2002. It was to comprise completely new vehicles, together with ground sensors, intelligent munitions and an integrating network. Beside four types of UAV, it foresaw three UGVs (with variants): the Armed Robotic Vehicle (ARV), the Small Unmanned Ground Vehicle (SUGV) and the Multifunctional Utility/Logistics and Equipment Vehicle (MULE). From 2005 on the General Accountability Office of the US Congress (GAO) noted considerable delays and cost overruns. In 2007 the Army cancelled three of the uninhabited vehicles. In 2008 the GAO warned that it is unclear if "the information network that is at the heart of the FCS concept can be developed, built and demonstrated"; in 2013 the Army will likely present "a partially developed and largely undemonstrated system for production" (GAO 2008, 2008a). After the change of the administration the FCS was halted in 2009; a much smaller Brigade Combat Team Modernization programme remains, with one armed UGV type (Army 2010). This shows that the ground-vehicle part of the Congressional mandate may be difficult to achieve.

In maritime vehicles, armed USVs are foreseen for anti-submarine as well as surface warfare, using guns, missiles or torpedoes (US Navy 2007). The Israeli firm Elbit offers the autonomous Silver Marlin of 11 m length, equipped with a sensor turret and a remotely controlled 7.62-mm weapon station (Elbit 2010). Armed UUVs exist already in the form of searching or fibre-guided torpedoes, development will probably go in the direction of increased autonomy, with use against submarines and surface vessels. The USA is developing new UUV types not only for finding sea and sea-floor mines, but also to destroy them.

In outer space, only first steps of arming uninhabited vehicles were done in the form of the anti-satellite systems of the 1970s and 1980s by the USSR and the USA; since then, a tacit moratorium holds. However, military visions of conquering outer space in a similar way as was done with air space have led to concepts of force application in and from space; such concepts were revived during the G.W. Bush administration. For attacking satellites, hit-to-kill interceptors could be used, or "servicing" satellites could manipulate others after rendezvous and docking.

In particular in the USA the military is intensely pressing ahead with research and development, co-ordinating and integrating among the services. Among the priorities for unmanned systems are: "Precision Target Location and Designation" and "Weaponization/Strike", for air, ground and maritime systems. The DoD states (2007: 54):

> "Weaponizing unmanned systems is a highly controversial issue that will require a patient "crawl-walk-run" approach as each application's reliability and performance is proved. ... Initial applications of weaponizing any unmanned systems may require a "man in the loop" ... to ensure positive control of the vehicle and its weapon. ... Guns, missiles, torpedoes, and nonlethal projectiles can "hang up" and create a potentially dangerous condition for unmanned systems recovery personnel and other platforms within the operating area. The challenge is the ability to remotely render unmanned weapon systems safe (with verification) or face the choice of having to destroy or scuttle the system. As confidence in system reliability, function, and targeting algorithms grows, more autonomous operations with weapons may be considered."

As primary technical challenges are mentioned: the ability to reliably target the right objective and achieve proper tracking under all conditions, and maintaining communications for man-in-the-loop operations.

Other countries follow the same route. Israel already has the Harop loitering UAV, which searches for a radar, flies into it and explodes. This UAV is the attack component of the German project WABEP,[6] to be used against more general targets; a second UAV, the German KZO, serves for observation and control (Rheinmetall 2010). Israel also deploys the larger UAVs Hermes 450 and Eitan/Heron TP in armed versions.

Iran possesses the smaller HESA Ababil-T UAV; it had transferred it to Hezbollah, which used it against Israel in 2004, 2005 and 2006 (only in 2006 they were shot down) (Gormley 2008: 65). In 2010, Iran announced two long-range UAVs equipped with bombs.

Because equipping indigenous or imported surveillance UAVs with weapons is not very difficult, one can expect a strongly increasing

---

[6] Wirkmittel zur abstandsfähigen Bekämpfung von Einzel- u. Punktzielen (Means for Attacking Single and Point Targets at Range).

number of countries commanding armed UAVs. Armed UVs on the ground and on or under water will follow markedly later.

5. Rules for Autonomous Killing?

For the time being the USA reserves the attack decision to a human operator. The DoD argues (2009: 10):

> "Because the DoD complies with the Law of Armed Conflict, there are many issues requiring resolution associated with employment of weapons by an unmanned system. For a significant period into the future, the decision to pull the trigger or launch a missile from an unmanned system will not be fully automated, but it will remain under the full control of a human operator.
> Many aspects of the firing sequence will be fully automated but the decision to fire will not likely be fully automated until legal, rules of engagement, and safety concerns have all been thoroughly examined and resolved."

However, there are several military motives to go in the direction of autonomous attack. One is technical advance – sensors and especially computers and software become more powerful. A second one is cost – if one soldier could control the operations of many robots, much personnel could be saved. Because humans are not good at handling many processes simultaneously, such an operator would only be able to execute general oversight, not controlling each individual attack. Consequently, there is already talk of moving from a human in the loop to a "human on the loop".

A further motive arises from the possibility that the communication link from the remote operator to the armed UV may be blocked – by overload of the transmission channel, by some malfunction, or by actions of the adversary, destroying or jamming communication nodes. In the future, when there would no longer be technological dominance of one side, not attacking when communication fails will mean giving the adversary an advantage. Finally, there is the issue of reaction time – if UVs of opponents meet each other, waiting for the satellite-communication delay plus human reaction time could result in loss of one's own UVs.

Given all these motives, it is not far-fetched to expect that – absent clear international prohibition – technology as well as procedures will glide down the slippery slope from „human in the loop" via „human on the loop" to „human far above the loop". The DoD states (2007: 53):

> "(...) the ultimate goal is to replace the operators with a mechanical facsimile or equal or superior thinking speed, memory capacity, and responses gained from training and experience."

Research in this direction has introduced several autonomy scales. One is fairly obvious, with four levels: remotely piloted (the human operator controls everything), remotely operated (the UV controls itself, with higher-order decisions by the human), remotely supervised (the UV fulfills its mission by itself, a human only intervenes if it does not carry it out correctly) and fully autonomous (a human sets general goals fulfilment of which is then planned and executed by the UV without human intervention). There is also a much refined scale with ten levels; here 1 means remote guidance, 4 applies to onboard route replanning, 7 is used for group tactical control, and 10 is reserved for fully autonomous swarms (OSD 2005: Fig. 5-5).

A few researchers have investigated the conditions and potential rules of engagement for autonomous weapons. John Canning of the US Navy proposes a concept of operations where the machines only target weapons, not humans (Canning 2009). Should the enemy soldier drop the weapon or leave the weapon carrier before the attack, his or her life would be spared. This would end war faster and bring lasting peace. In case a human needs to be killed directly, the armed vehicle should reduce its autonomy and transfer control to a remote human operator.

Philosopher Patrick Lin and coworkers at the California Polytechnic State University (USA) find that "[c]reating autonomous military robots that can act at least as ethically as human soldiers appears to be a sensible goal" (Lin at al. 2008).[7] However, they point out that there are daunting challenges, in particular with discrimination, and call for more interdisciplinary research concerning issues of risks and ethics.

Ron Arkin (2009),[8] robotics researcher at the Georgia Institute of Technology (USA), thinks the "robots not only can be better than soldiers in conducting warfare in certain circumstances, but they also can be more humane in the battlefield than humans." He points out

---

[7] This project was funded by the US Office of Naval Research.
[8] This project was funded by the US Army Research Office.

that "robots can be built that do not exhibit fear, anger, frustration, or revenge" and would not need a strong self-preservation drive. He proposes to program the laws of war – guided by the principles of military necessity, avoidance of unnecessary suffering, proportionality between damage produced and military advantage gained, and discrimination between combatants and non-combatants – into the algorithms controlling the autonomous weapon. Beside ethical behaviour control there is an ethical governor (to do a second check before a lethal act); its constraints are to be adapted after action by reflection, or by affective functions (guilt, remorse, grief) if a violation of the laws of war occurs. There should be a right to refuse an order which the system finds unethical, but with the possibility of an explicit override by a human operator together with a second officer. Also Arkin speaks of daunting problems and stresses the need for future research.

The main question here is: Will the forseeable „intelligence" of artificial-intelligence systems suffice to warrant judgement and action at least on the level of human capabilities? This is what international humatarian law demands. Arkin's concept has been criticised on several grounds. P. Asaro (2009), philosopher at the New School University, New York (USA), contends that the laws of war, the respective rules of engagement for an operation and just-war theory "cannot be easily realized within an automatic system because they are actually a hodgepodge of laws, rules, heuristics, and principles, all subject to interpretation and value judgments." Noel Sharkey (2007), robotics researcher at the University of Sheffield (UK), points to the complexity in a military operation where there are civilians too. "Many different events can occur simultaneously, giving rise to unpredictable or chaotic robot behavior." Also, behaviour compatible with the proportionality requirement would need quantifying the military advantage and the amount of collateral damage, including killing of children (Sharkey 2009). "Ultimately, we must ask if we are ready to leave life- or death decisions to robots too dim to be called stupid." (Sharkey 2007)

6. Assessment of Armed Uninhabited Vehicles under Criteria of Preventive Arms Control

Throughout the history of war, technological superiority has been an important factor in who gained the upper hand. Since the second World War, the industrialised countries have used science and technology systematically to make their armed forces more effective and more efficient. However, the efforts to achieve national security by strong armed forces suffer from a fundamental problem, the so-called security dilemma, following from the basically still anarchic character of the international system. Usually, stronger armed forces also mean a higher potential for attacking others, increasing the mutual threat. Thus the net result of the countries' quest for their own security is a reduced security for all. This is evident with the various steps in the build-up of nuclear forces in the Cold War (hydrogen bomb, long-range ballistic missile, multiple independently targetable reentry vehicles, leading to shorter reaction times and increased pressure to act fast). However, the same mechanism is valid in other areas of military competition, also after the Cold War. One of the ways out of the security dilemma is arms control in general – limiting numbers or qualities of weapons and armed forces by mutual agreement, with adequate verification of compliance.

Concerning new military technologies states can counteract the deterioration which may ensue by preventive arms control, that is limiting or prohibiting certain potential new military systems before they are acquired and deployed (Altmann, 2006: Ch. 5). Some existing arms-control agreements contain prohibitions already for the stages of development and testing, for example, the Biological Weapons Convention of 1972 and the Chemical Weapons Convention of 1993. An example from international humanitarian law is the Protocol banning blinding laser weapons of 1995 which prohibits only their use, but had the effect that their development and even research was practically stopped.

Preventive limitations can only be concluded among states – criminals and terrorist groups cannot be parties. However, such groups have only very limited capabilities of doing research and development on their own. Regulation binding the states can go a long way in preventing access by these groups to technologies and systems which, absent limitations, would be developed in many countries and proliferate, on open or black markets.

When investigating whether military applications of a new technology should be limited, one can use the criteria of preventive arms control. They have been categorised in three groups:

I. Adherence to and further development of effective arms control, disarmament and international law
   1. Prevent dangers to existing or intended arms-control and disarmament treaties
   2. Observe existing norms of international humanitarian law
   3. No utility for weapons of mass destruction
II. Maintain and improve stability
   1. Prevent destabilisation of the military situation
   2. Prevent technological arms race
   3. Prevent horizontal or vertical proliferation/diffusion of military-relevant technologies, substances or knowledge
III. Protect humans, environment and society
   1. Prevent dangers to humans
   2. Prevent dangers to the environment and sustainable development
   3. Prevent dangers to the development of societal and political systems
   4. Prevent dangers to the societal infrastructure.

Groups I and II are about the prevention of armed conflict and how it is waged, group III concerns dangers arising in peacetime, such as new pollutants, or systems that could proliferate to terrorists.
A systematic analysis of the various potential types of armed uninhabited vehicles under these criteria still needs to be done. A first short assessment arrives at the following preliminary conclusions (Altmann 2009, see also Sparrow 2009 and Altmann 2006):

Arms Control and Disarmament (I.1)
Due to the general-purpose criteria used in the Biological Wesapons Convention of 1972 and the Chemical Weapons Convention of 1993, uninhabited vehicles must not be equipped with such agents or release mechanisms for them. Dangers would only ensue if there were intentions to do so.
Concerning nuclear weapons, the New Strategic Arms Reduction Treaty between the USA and Russia (New START, 2010) allows new

types of arms. Insofar as new uninhabited combat air vehicles could be equipped with nuclear bombs or missiles, they could undermine the New START rules on cruise missiles and bombers. The same holds for hypervelocity missiles and transatmospheric vehicles with respect to ballistic missiles, because the former would not fly ballistically over most of their trajectory. The Intermediate-range Nuclear Forces Treaty of 1987 between the USA and the USSR/Russia prohibits ground-launched ballistic missiles as well as ground-launched cruise missiles with ranges between 500 and 5,500 km (air- and sea-/submarine-launched cruise missiles are not affected). If uninhabited aircraft with a range in this interval would be equipped with nuclear weapons, the respective country would certainly argue that these UAVs would not constitute cruise missiles and would thus not be constrained – this may be relevant because the US DoD definition of unmanned vehicle excludes cruise missiles (see Section 2). In this way, the Treaty could be undermined. An additional problem is that there is no comparable treaty binding other countries.

The Treaty on Conventional Armed Forces in Europe of 1990, unfortunately suspended at present, limits the holdings in five major weapons classes for the NATO member states and Russia. Intentionally its definitions of battle tanks, armoured combat vehicles, artillery, combat aircraft and attack helicopters do not mention personnel on board. Thus for uninhabited versions the same rules would apply. However, one can foresee arguments that armed UAVs smaller than traditional combat aircraft or attack helicopters should count differently, and discussions which portion of multi-purpose UAVs should count can become very complicated. Different from aircraft, "battle tanks" and "heavy armament combat vehicles" are defined using mass thresholds (16.5 tons and 6.0 tons, respectively) so that potential future uninhabited combat ground vehicles with lower masses would fall outside of these traditional categories, opening a grey area of uncounted and unlimited combat systems. A fundamental problem is that in other continents than Europe there are no comparable limitations of conventional armaments.

The Outer Space Treaty of 1967 prohibits nuclear weapons and other weapons of mass destruction in Earth orbit, whether or not a crew is present.[9] However, other weapons are not prohibited, despite periodic

---

[9] Ballistic missiles do not count because they travel through outer space only temporarily.

votes of the UN General Assembly for the prevention of an arms race in outer space (e. g. United Nations, 2008). Plans for docking, servicing and manipulation satellites without crew open new possibilities for attacks on satellites, endangering progress towards the long-sought-for general ban of space weapons.

International Humanitarian Law (I.2)
Armed UVs can malfunction; the obligation to protect non-combatants can mean to destroy one if control is lost, as done in Afghanistan in September 2009, when a Reaper UAV was shot down by a piloted combat aircraft of the US Air Force (US Air Force 2009).

UVs, in particular UAVs, provide the possibility for better discrimination. Compared with artillery and aerial bombing their attacks can be better directed and thus reduce collateral damage. Also, they allow real-time judgement of and reaction to changes on the scene, but soldiers on ground can do both better and have more possibilities to act.

The armed-UAV attacks by the USA in Afghanistan, Iraq and Pakistan show that in many cases the principle of discrimination in warfare is violated – civilians are attacked and killed (Weber 2009, New America Foundation 2011). Because video images are the main information provided to the remote operator, misjudgement is easily possible. Even helicopter crews observing a scene directly can be wrong, as drastically demonstrated in July 2007 when two Reuters correspondents with more than ten other civilians were killed in Baghdad; it seems that their tele-lens photocameras were mistaken as rocket launchers or assault rifles (WikiLeaks 2010).

In addition to limited information available, in case of remote control from the home country the office environment and computer-game-like interface make it easier to press the button (Cummings 2004; Weber 2009), but the distance effect applies also to aircraft pilots. A UV – at least a UAV – has shooting as its only potential action, different from a soldier on the ground who could check the identity, search for weapons and arrest a person.

In case of autonomous attack decisions there are more fundamental issues. One is whether killing of humans by a machine should be allowed at all. Then there is the requirement of judgement at the same level as that of a human commander: discriminate between a combatant and a non-combatant, recognise when a combatant is *hors de combat* (for example unconscious or surrendering), balance military advantage against collateral damage. Judging on very complex, fast-varying situations at the human level is far from the present state of "artificial intelligence" and might not be achieved even after several decades.

Weapons of Mass Destruction (I.3)
UVs can carry weapons of mass destruction. As mentioned above, biological and chemical weapons fall under comprehensive prohibitions, but nuclear weapons are not (yet) outlawed. Nuclear weapons on board satellites are banned (see I.1 above). Nuclear-armed submarines and surface ships will probably remain to be operated by a crew. Land vehicles carrying a nuclear weapon to a target are not attractive for the military – they might be seized and would take considerable time for longer distances. UAVs are the most likely nuclear-weapon carrier. Cruise missiles exist already, but uninhabited nuclear bombers are conceivable, as are smaller re-usable aircraft.

Destabilisation of the Military Situation (II.1)
Destabilisation occurs when military motives to start a war or to attack become stronger, in particular if one fears a significant disadvantage in case one waits for an adversary to attack. In this respect some UAVs are problematic because they can penetrate deeply into enemy territory to carry out precision surprise attacks. At low altitude and relatively slow speed they are very difficult to detect and defend against. With no humans on board, UAVs could be sent more easily and for more dangerous missions. Destabilisation could also result if the attacked side would be uncertain if the vehicles carry weapons of mass destruction.

Destabilisation is particularly relevant in a crisis. Pressures to act fast would build up whenever uninhabited combat vehicles of two potential opponents would meet at short distance, along a border or in international territory. Two fleets of armed UAVs would watch each other, on high alert. A co-ordinated attack by one side could wipe out a significant portion of the systems of the other. Any erroneous sign of attack could trigger actual shooting, leading to war by uncontrolled feed-back cycles between the two systems of warning and attack. Delays from remote control may endanger one's own combat systems, creating a strong motive to delegate firing authority to the systems.

Instability and nervousness can result from small satellites which could dock and manipulate important satellites of an opponent, potentially destroying on short notice the capability for strategic warning, surveillance and communication. Swarms of highly precise conventionally armed small UAVs might even be used to disrupt nuclear-strategic targets, creating instability at the level with the highest consequences. More into the future, micro-robots could enter military systems of a potential opponent covertly, sitting there unnoticed, ready to disrupt the electronics at any time.

Technological Arms Race (II.2)
A qualitative arms race can already be observed at present with unarmed UAVs. But the number of countries developing and deploying armed UAVs is also rising. Technologies developed and deployed by leading countries will be introduced by others with only a few years delay. Measures taken by one country will lead to countermeasures by others, and so on. Similar developments over time are probable with maritime and land vehicles.

Horizontal or Vertical Proliferation (II.3)
Horizontal proliferation – that is, to other countries – of technologies and systems is already taking place with unarmed UAVs – over 50 countries use, produce or develop them and 20 countries export them. The few countries possessing armed UAVs today will soon be joined by several others. In the future, the world will see more exports, including to crisis regions. Horizontal proliferation can also occur by collaborative development.
Vertical proliferation, the qualitative improvement of military technologies or systems, is evident with UAVs in many countries. The next step of arming them is underway, to be followed by UAVs capable of all air-power missions, including bombing or air combat. Depending on development success, horizontal and vertical proliferation can arise with UVs in other media, too.

Dangers to Humans (III.1)
If military armed UVs will proliferate widely, some of them could find their way into the hands of criminals. Small UAVs would make ideal terrorist tools, carrying tens of grams to several kilograms of payload. In the form of explosive, only limited effects would be

produced; strong indirect effects are nevertheless possible by attacking important persons, even if only mechanical impact were used. Mass destruction could be achieved by chemical and in particular biological agents. For some scenarios, water or land systems could also play a role.

Such dangers would be higher if military UVs will have been designed for such purposes. On the other hand, many UVs including civilian, unarmed ones will be modular, useable for many different payloads.

Dangers to Environment and Sustainable Development (III.2)
This area is not very relevant here.

Dangers to Societal and Political Systems (III.3)
Societal and political systems could be endangered by UVs used for terrorism, as discussed in III.1. Furthermore UVs could be used for reconnaissance for the preparation of terrorist or other criminal attacks by other means.

Dangers to Societal Infrastructure (III.4)
Terrorists could use UVs with many kilograms of payload to directly damage or destroy infrastructure installations. Smaller systems could hit central control and communication components, or provide intelligence for later attacks by larger systems or a group of people.

These considerations show that there are strong reasons for concern about armed UVs. Small ground vehicles remotely controlled over tens of metres for inspection and destruction of roadside bombs seem harmless at first sight, but could become dangerous if they will be armed and controlled from far away or autonomously. On the international level, the biggest problems arise with respect to the laws of war and from destabilisation and proliferation. This holds for military systems of all sizes. Within societies, the most important issues arise from smaller, cheaper, more easily accessible systems that would be used by terrorists or other criminals. As a consequence, limitations by preventive arms control are advisable.

Some recognition of such dangers by states can be seen in export-control regimes. Different from most arms-control treaties these regimes are discriminatory and not legally binding. The Missile Technology Control Regime restricts exports of ballistic missiles as well as cruise missiles (and their components), which can carry 500 kg or more over 300 km or more. For UAVs there is no payload threshold, and if they have autonomous or long-range remote-control navigation and aerosol-spraying equipment, there is no range threshold either. The Hague Code of Conduct asks for restraint in exports of ballistic missiles only. The Wassenaar Agreement restricts exports of armament technologies and sensitive dual-use technologies many of which are relevant for uninhabited vehicles. While these measures have a limiting effect on horizontal proliferation, export controls suffer from the principal problems that not all countries participate and that the possessor countries do not restrict their own use and further development of the respective systems and technologies.

7. Options for Preventive Arms Control

Developing concepts for preventive limits for armed UVs is a major task and needs considerable international debate. At the present state, several general considerations can be given, followed by options for limitation (Altmann 2009, see also Sparrow 2009, ICRAC 2010).

Because UVs can be used for several beneficial purposes, there should not be a complete prohibition. In order to prevent circumvention of limits on military UVs by civilian ones, the latter should be included in international regulation and verification in some way. To keep the intrusiveness of verification limited and rely on-site inspections without special equipment mostly, there should be no UVs/very small robots below 0.2 to 0.5 m size, with narrowly circumscribed exceptions for example for the exploration of shattered buildings. Bigger, unarmed UVs for military or civilian purposes should be regulated and limited numerically. There should be no new categories of UVs carrying nuclear weapons beside the existing ballistic and cruise missiles, which should be reduced in the re-started process of nuclear disarmament. Uninhabited satellites with weapons or used as weapons should be prohibited in the framework of a general ban on space weapons; docking and manipulation satellites should be regulated internationally.

Concerning limitation options, optimally armed UVs would be banned outright, except for the types already deployed before 2001, such as ballistic and cruise missiles, standoff missiles, missiles for air defence and defence against ballistic missiles, anti-ship missiles, torpedoes. This would mean withdrawal of a few new types deployed after 2000, mainly by the USA, such as Predator and Reaper. For most other countries such a prohibition would affect only future systems. To be comprehensive and increase the barrier against covert circumvention, the prohibition should cover the phases of development, testing, deployment and use. This would stop the next wave of new classes of combat aircraft/land/sea vehicles from the beginning.

For the armed-UV types existing before 2000, limits along the lines of the Treaty on Intermediate-range Nuclear Forces or the Treaty on Conventional Armed Forces in Europe should be sought after worldwide.

If a general ban on new armed UVs cannot be achieved, then a set of specific rules and limitations should be agreed upon internationally. International humanitarian law should be augmented by an explicit prohibition of autonomous machine decision on whom or what to attack, absolutely demanding a human for each individual weapon release.[10] To contain destabilisation, various tools are available. Depending on UV category, maximum holdings and diffentiated limits on range, endurance, payload can be introduced. Geographical criteria can be used, in particular deployment close to other countries can be prohibited or limited. To alleviate verification and to inhibit UV use by terrorists, UVs below a certain size (0.2 to 0.5 m) can be banned (except in very low numbers with full transparency for accepted civilian purposes). Additional detailed rules could be developed.

8. Outlook

Up to now, the development of armed UVs has followed the logic of one's own military strength. However, via interactions in the international system this will likely lead to arms races, destabilisation and increased terrorist threats. Thus, in the interest of peace and international stability preventive arms control is advisable. To promote the debate about the necessity and design of limits on armed UVs and to support a political process towards negotiations, a few scientists from robotics, ethics and peace research have founded an

---

[10] Existing target-seeking missiles with simple target classes (such as radars or aircraft) can be exempted by a list of such types.

International Committee for Robot Arms Control in 2009 (ICRAC 2011).[11] Roboticists and artificial-intelligence researchers should be aware of the military uses and their dangers and should support international limits.

References

Alston, P. (2010): *Report of the Special Rapporteur on extrajudicial, summary or arbitrary executions, Philip Alston – Addendum – Study on targeted killings*, United Nations General Assembly, Human Rights Council, A/HRC/14/24/Add.6, 28 May, http://www2.ohchr.org/english/bodies/hrcouncil/docs/14session/A.HRC.14.24.Add6.pdf (30 Sept. 2010).

Altmann, J. (2006): *Military Nanotechnology: Potential Applications and Preventive Arms Control*, Routledge, Abingdon/New York.

Altmann, J. (2009): "Preventive Arms Control for Uninhabited Military Vehicles", in: Capurro, R. & Nagenborg, M. (eds.): *Ethics and Robotics*, AKA, Heidelberg.

Arkin, R. C. (2009): *Governing Lethal Behavior in Autonomous Robots*, Chapman&Hall/CRC, Boca Raton FL.

Army (2010): *Army Brigade Combat Team Modernization – Capabilities Overview*. Available at: https://www.fcs.army.mil/systems/index.html (9 July 2010).

Asaro, P. (2009): "Modeling the moral user", *IEEE Technology and Society Magazine*, Vol. 28, No. 1, pp. 20–24.

Canning, J. (2009): ""You've just been disarmed. Have a nice day!"", *IEEE Technology and Society Magazine*, Vol. 28, No. 1, pp. 13–15.

Cummings, M.L. (2004): "Creating Moral Buffers in Weapon Control Interface Design", *IEEE Technology and Society Magazine*, Vol. 23, No. 3, pp. 28–33 & 41.

DoD (Department of Defense) (2007): "UMS Roadmap 2007-2032", Washington DC, US Department of Defense. Available at: http://www.acq.osd.mil/usd/Unmanned%20Systems%20Roadmap.2007-2032.pdf (30 Juli 2008).

DoD (Department of Defense) (2009): "FY2009–2034 Unmanned Systems Integrated Roadmap", Washington DC, Department of

---

[11] The international expert workshop organised in Berlin by ICRAC in 2010 demanded a ban on autonomous killing and limitations on remotely controlled armed UVs (ICRAC 2010).

Defense. Available at: http://www.jointrobotics.com/documents/library/UMS%20Integrated%20Roadmap%202009.pdf (16 April 2010).

Drew, C. (2009): "Drones Are Weapons of Choice in Fighting Qaeda", *New York Times,* March 16.

Elbit Systems (2010): *Unmanned Surface Vessel*, Available at: http://www.elbitsystems.com/lobmainpage.asp?id=1025 (8 July 2010).

Gage, D. W. (1995): "UGV History 101: A Brief History of Unmanned Ground Vehicle (UGV) Development Efforts", *Unmanned Systems Magazine*, Vol. 13, No. 3.

GAO (Government Accountability Office) (2008): *Defense Acquisitions – 2009 Is a Critical Juncture for the Army's Future Combat System*, US Government Accountability Office, Washington DC, March.

GAO (Government Accountability Office) (2008a): *Defense Acquisitions – Significant Challenges Ahead in Developing and Demonstrating Future Combat System's Network and Software*, US Government Accountability Office, Washington DC, March.

Gormley, D.M. (2008): *Missile Contagion – Cruise Missile Proliferation and the Threat to International Security*, Annapolis MD, Naval Institute Press.

Hambling, D. (2007): "Military Builds Robotic Insects", *Wired*, January 23. Available at: http://www.wired.com/science/discoveries/news/2007/01/72543 (7 July 2010).

Hambling, D. (2010): "Air Force Completes Killer Micro-Drone Project", *Wired – Danger Room*, 5 January. Available at: http://www.wired.com/dangerroom/2010/01/killer-micro-drone/ (7 July 2010).

Hersh, S.M. (2002): "Manhunt – The Bush Administration's new strategy in the war against terrorism", *The New Yorker*, December 23.

Hoyle, C. & Koch, A. (2002): "Yemen drone strike: just the start?", *Jane's Defence Weekly*, Vol 38, No. 20, p. 3.

ICRAC (2010): *Statement of the 2010 Expert Workshop on Limiting Armed Tele-Operated and Autonomous Systems, Berlin, 22nd September*. Available at: http://www.icrac.co.uk/ Expert%20Workshop%20Statement.pdf (9 May 2010).

ICRAC (2011): *International Committee for Robot Arms Control*, Available at: http://www.icrac.co.uk, May 9.

Jane's (2007): *Jane's Unmanned Vehicles and Aerial Targets*, Coulsdon, Jane's.

Jewell, L. (2004): "Armed Robots to March into Battle", *Army News Service*, 6 Dec. Available at: http://www.defense.gov/transformation/articles/2004-12/ta120604c.html (7 July 2010).

Krishnan, A. (2009): *Killer Robots – Legality and Ethicality of Autonomous Weapons*. Farnham Surrey/Burlington VT, Ashgate.

Lin, P.; Bekey, G. & Abney, K. (2008): *Autonomous Military Robotics: Risk, Ethics, and Design*. San Luis Obispo: California Polytechnic State University, Ethics & Emerging Sciences Group, December. Available at: http://ethics.calpoly.edu/ONR_report.pdf (1 Febr. 2010).

Mayer, J. (2009): "The Predator War – What are the risks of the C.I.A.'s covert drone program?", *The New Yorker*, October 26.

New America Foundation (2011): *Counterterrorism Strategy Initative – The Year of the Drone*. Available at: http://counterterrorism.newamerica.net/drones (9 May 2011).

Newcome, L.R. (2004): *Unmanned Aviation: A Brief History of Unmanned Aerial Vehicles*, American Institute of Aeronautics and Astronautics, Reston VA.

OSD (Office of the Secretary of Defense) (2005): *UAS Roadmap 2005-2030*. Department of Defense, OSD (Office of the Secretary of Defense), Washington DC.

Qinetiq North America/Foster Miller (2010): *TALON Family of Military, Tactical, EOD, MAARS, CBRNE, Hazmat, SWAT and Dragon Runner Robots*. Available at: http://foster-miller.qinetiq-na.com/lemming.htm (7 July 2010).

Rheinmetall Defence (2010): *Wabep Weapon System*. Available at: http://www.rheinmetall-detec.com/index.php?fid=4251&lang=3&pdb=1 (8 July 2010).

Sanger, D.E. & Shanker, T. (2010): "U.S. Faces Choice on New Weapons for Fast Strikes", *New York Times*, April 23.

Sharkey, N. (2007): "Automated Killers and the Computing Profession", *IEEE Computer*, Vol. 40, No. 11, pp. 124, 122–123.

Sharkey, N. (2009): "Death strikes from the sky: the calculus of proportionality", *IEEE Technology and Society Magazine*, Vol. 28, No. 1, pp. 16-19.

Singer, P. (2009): *Wired For War – The Robotics Revolution and Conflict in the 21st Century*, Penguin, New York.

Sparrow, R. (2009): "Predators or plowshares? Arms control of robotic weapons", *IEEE Technology and Society Magazine*, Vol. 28, No. 1, pp. 25–29.

United Nations (2008): "Prevention of an arms race in outer space", *UN General Assembly Resolution, 63th session, A/RES/63/40, Agenda Item 88*, December 2. Available at: http://www.un.org/ga/63/resolutions.shtml (16 December 2008).

US Air Force (2009): "Reaper crashes in Afghanistan", *News*, 14 September. Available at: http://www.af.mil/news/story.asp?id=123167644 (22 November 2009).

US Navy (2007): *The Navy Unmanned Surface Vehicle (USV) Master Plan*. Department of the Navy, Washington DC, July. Available at: http://www.navy.mil/navydata/technology/usvmppr.pdf (8 July 2010).

van Blyenburgh, P. (ed.) (2007): *UAS – Unmanned Aircraft Systems – The Global Perspective 2007/2008*, Blyenburgh & Co, Paris.

Weber, J. (2009): "Robotic Warfare, Human Rights & the Rhetorics of Ethical Machines", in: Capurro, R. & Nagenborg, M. (eds.): *Ethics and Robotics*, AKA, Heidelberg.

WikiLeaks (2010): *Collateral Murder*. Available at: http://www.collateralmurder.com (12 May 2010).

Investigating the Robot in the Loop.
Technology Assessment in the Interdisciplinary Research Field Service Robotics

Martin Meister

Abstract: The basic challenge of robotics is characterized as facing the "human frontier". Smart robots entering the households and workplaces of ordinary people is without any doubt a societal issue of great impact, which calls for a detailed technology assessment. But such an assessment is not easy, for two reasons. First, there are two completely different versions or readings of the notion of the ihuman frontierî. For engineering and Artificial Intelligence, human environments are the most demanding settings for a robot to operate. For the Social Sciences, and especially for the emerging field of Roboethics, these robots are seen as a major threat. The article sketches these two opposing viewpoints and advocates a more balanced view. Second, technology assessment faces the problem of finding a reliable source of information and the right target for their recommendations, because the "New Robotics" is a massively heterogeneous field of research. By following some of the coordination practices in the field itself the article proposes some possible starting points for alignments with different strands of research, with the focus on the possibilities of investigations of human-robot interactivity in realized scenarios.
Keywords: Service Robots, Technology Assessment, Human-Robot-Interaction, heterogeneous Fields of R&D

In the recent handbook of robotics research, which gives an encompassing overview of the many strands of robotics research, the editors summarize the basic challenge of today's robotics as "the human frontier":

> "Reaching for the human frontier, robotics is vigorously engaged in the growing challenges of new emerging domains. Interacting, exploring, and working with humans, the new generation of robots will increasingly touch people and their lives. The credible prospect of practical robots among humans is the result of the scientific endeavor of a half a century of robotic developments that established robotics as a modern scientific discipline." (Siciliano & Khatib 2008: XVII)

As is noticeable, this statement is not formulated as a vision for the future. It summarizes the belief of many researchers, that the 'New Robotics' – autonomous mobile machines that are capable of navigating in and adapting to changing environments, and doing reasonable things in this environment – has developed to the point where these machines can enter the sphere of human everyday life. Therefore, it is no big surprise that human-robot-interaction is a big issue in the Handbook.

Smart robots entering our households and workplaces, and the homes of seniors and disabled, as usual the sites of the most demanding applications, as guides and guards and partners, is without any doubt a societal issue of great impact. And these technical developments have boosted a broad debate about Roboethics, an issue that is also included in the Handbook, prominently: as the concluding chapter. So this development seems to call for a detailed technology assessment (TA), and I agree. But such a TA is not an easy thing to do, for two reasons: First, there are two completely different versions or readings of the notion of the "human frontier", one from engineering and Artificial Intelligence (AI), and the other from the social sciences. I will try to sketch a more balanced view. And second, TA as well as roboethics face the problem of finding a reliable source of information for their investigations and assessments, and the right target for their recommendations, because the 'New Robotics' is a massively heterogeneous field of research. By following some of the coordination practices in the field itself I will try to identify possible alignments of different strands of research.

1. Two Versions of "The Human Frontier"

From the robotics researchers' point of view, regarding their visions, concepts and laboratory practices, human environments are the most demanding settings for a robot to operate – and thus the ultimate challenge. Other than industrial robots, which repeatedly do the same things in an accurately defined surrounding, and other than field robots, which operate far away from humans, service robots are thought to operate in the presence of the most disturbing and unpredictable elements you can imagine: ordinary human beings.[1]

---

[1] Dividing robotics research in these three strands is common in the field, with service robotics as the newest (but historically oldest) and most challenging part of the robotics endeavour (see cf. Kawamura et. al. 1996). All three areas have their own conference series, journals, market leaders for equipment, and so on. Besides this basic division, there are at least three other strands of robotics research and application: Robotics in entertainment, in arts, and intelligent extensions of the

Humans, by purpose or simply by habit, do not follow any formally defined rules, they permanently change the environment, they gather to crowds or want to be alone, they shout or mumble, and so on. All of these everyday human activitities bring out tremendous challenges for a robot, concerning self localization and navigation, steering model and decision making, sensors and interface design, to name but a few of the technical difficulties that have to be solved. Moreover, all these single difficulties have to be integrated in one platform, which should allow a real-time coordination of different hardware and software components on one platform, which is the basic challenge in robotics[2]. To create a robot that is capable to operate in such a "complex" environment, the humans in this environment have to be taken into account, at least somehow. Some engineers think this is only a question of finding the right "business case", which will convince the audience of the opportunities which are connected to this class of clever robots, esp. in economic terms (see, c.f., Prassler&Kosuge 2008). But as the Handbook also shows, in many approaches the issue of man-robot interaction is increasingly regarded as part of the inner core of robotics research itself. This issue, however, is conceptualized quite differently. As in every technically driven research, one approach is to improve the interfaces of the robots, especially to make them more "human-like" and thus "intuitively" usable, and to apply multi-modal interfaces, including not only typing and speech recognition, but also recognition of gestures and even emotions (Breazeal et. al. 2008). Another very common approach is to try to understand and explicitly model human behavior and reasoning so comprehensively that the model grasps its most relevant aspects, then to formalize it as algorithms and architecture and to transfer this formal model to the robots architecture and reasoning in order to become more human-like (or 'natural'). This approach of course echoes the old dream of AI, now in its embedded and embodied version. Or the hard to grasp behavior of he humans is used as a source for improving robots performance, as in "Learning by Demonstration" where the human is the 'teacher' and the robot the

---

human body: Intelligent exoskeletons for soldiers (or disabled people), and intelligent prostheses (mainly for disabled people).

[2] See cf. Murphy 2000 for an encompassing introduction, which is readable for non-engineers, to the issue of structured integration of an intelligent robots components in an architecture which couples the "sense-think-act"-scheme of every robot.

'pupil' for a specific task, which again is formally modeled as a (human as a mechanical) "skill" (Billard et. al. 2008).

Despite their differences, in all of these approaches the ultimate goal is to improve the performance of the robots and – with the possible exception of approaches inspired by Cybernetics, on which I will turn later on – human everyday activity is nothing but a source for understanding and formally describing the most challenging surrounding for an advanced robot. With respect to human-robot-interaction, the underlying assumption is that a robot that resembles a human in appearance, behavior and interface will be more easily accepted by users. And with respect to the specification of the "new emerging domains" cited above, that is new areas of application, almost all of the researchers follow the "I-Methodology" (as described by Akrich 1995), which means that there is no empirical investigation in the needs, visions and actual usages of ordinary people at all – developers simply imagine themselves as the users.

Whereas the possibilities of the "New Robotics" are formulated from the developers' standpoint as a challenge for the machines, from a TA-standpoint the "human frontier" by nature is formulated the other way round. Autonomous robots are perceived as crossing the red line to a new realm, our everyday life – which is without any doubt a major threat, as long as the possible consequences for humans and society at large are not thoroughly assessed. To make this point is not professional scrupulosity (or "German Angst"), because everyone who knows robotics researchers knows that they love their creatures (a bit like gadgets, more as pets, but even more and quite literal like children) and simply cannot imagine that these creatures, with all their hard to achieved behaviors, could do ordinary humans any harm.

The debate about the social impacts of "New Robotics" is dominated by the discourse of "roboethics", which comes in very different versions, depending on the estimation about robotics achievements in the near future. There is much talk about implanting moral principles to autonomous robots (it would be very interesting to see a Kant-module working in an operating robot), which echoes Asimov's famous laws for robots from the fifties[3], or about the possibility to ascribe full fledged moral responsibility to advanced robots.

The authors of the chapter in the Handbook about Roboethics (Veruggio & Operto 2008) choose a more down-to-earth approach.

---

[3] Asimov's laws look, despite being pure science fiction, a lot more adequate.

They are skeptical about the grand visions (and the grand fears) attached to advanced robotics, or even about the discourse of the evolution of a new species. Instead, they are taking as the starting point that even very advanced robots are still machines programmed by someone for a specific task, and not a potentially ethical responsible artificial being. The targets of roboethics, then, are not the robots themselves, but the scientists and engineers who design and develop them. Consequently, the authors and their colleagues try to establish roboethics not as an endeavor distanced from the field of research and development, but as part of it – as the processor developing a "professional responsibility" that should belong to the field itself (ibid.: 1510). The mission of roboethics is formulated accordingly as follows:

> "Roboethics is not a veto or a prohibitionist ethics. Its main lines of development are: the promotion of culture and information; the permanent education; a vigorous and straight public debate; and the involvement in all these activities of the young generations who are the actors of the future." (ibid.: 1515)

The proposed way of fostering and implementing roboethics within research and development is not the demand for restrictive measures or even laws, but to raise consciousness – by jointly formulating roadmaps (like the EURON roboethics roadmap in the EU), by formulating and systematizing guiding principles for professional practice in engineering, computer science and AI, by implementing it into curricula for students of the disciplines involved, and by stimulating public debate.

By doing this, roboethics seems to do a good job as a TA for the "New Robotics", not only because of identifying its possible impacts and treats, but also because its approach seems to echo the principles of newer version of TA like "Constructive Technology Assessment" (in the EU; Schot & Rip 1997) or "Real-time Technology Assessment" (the term in the US for the same idea; Guston & Sarewitz 2002). Though not explicitly mentioned, this seems in line with the request to "modulate" (Schot & Rip 1997) the process, which means to constructively influence the technical development in its early stages instead of any request for prohibition of certain technical developments or applications.

With regard to contents, a broad range of possibly harmful impacts of the "New Robotics" for humans are identified and discussed with the goal to create a systematic taxonomy, not at least with respect to the specific criteria for different application areas of robotics. This list of possible negative impacts is large, but I think it can be subdivided into three main parts.

- There are the cases with undeniable evidence of dangers, which have to do with the physical proximity of the robot to humans. A rehabilitation robot or a mobile household helper for seniors, for example, must be constructed in a way that hurting the entrusted humans by bumping into them with body or arm is definitely impossible. The authors do not give this issue great attention and treat it as a somehow trivial matter of course. But as the respective chapters of the Handbook show, developers, though they are very aware of this type of problems, perceive the issue as a limitation. They equip the robots with – typically more than one – security installation for the unpredictable behavior of humans in order to protect these[4], which in their view is a necessary compromise because it encapsulates the robots intelligence from some aspects of its environment and thus contradicts the ultimate goal to achieve autonomous behavior[5].

- Then the list includes a lot of issues well established from ethics of information technology, like privacy, reliability and dependability, dual use (esp. for military applications),

---

[4] Typically, this is done by proximity sensors, which stop robots movements at all if a human is approached to near, and by installing an easy to use 'big red' stop-button, that is also part of the typical laboratory settings.

[5] The episode I most like happened in a robotics exebition in the German Telecommunication Museum in Berlin two years ago. When strolling through the exebition, suddenly a robot approached at high speed, stopping just before us (the proximity sensors) yelling: "Hello, I am Robbie, can I help you?" Just behind this robot came two human exebition guards hurrying, saying "Don't worry, it does nothing" (like dog-owners) and explaining, that the robot went defunct because of children playing with it, and they have to wait for the original programmers to return and fix it. Meanwhile, the museum guard had a double task: To look after the visitors, and to guard the defunct robot, explaining the visitors, that this machine is advanced, but at the moment defunct.

> replacement of humans, or digital divide. These issues are without any doubt relevant, but not specific for robots[6].

- And finally the authors discuss some issues, which address this specificity: the expectation that "we are going to be cohabiting with robots endowed with self-knowledge and autonomy" (Veruggio & Operto 2008: 1511). Possible negative impacts are formulated as "psychological problems" like "deviations in human emotions, problems of attachment … fears, panic … feeling of subordination towards robots" (ibid.: 1512). Many if not all of these issues are presented as consequences of the likeliness of a "confusion between the real and the artificial" (ibid.). I will turn to this point later on.

Though worthwhile for systematizing a broad range of issues, and for establishing assessment in the field of research and development, roboethics in my view also has three major shortcomings. First, its implementation strategy is solely based on discourse, professional and public, and creating guidelines. This does not address the core practices of the engineering and design processes, especially in a massively heterogeneous field. Second, roboethics is presented not only as an ethics for robot developers, but restricted to their perspective in the sense of the aforementioned "I-methodology", quite explicitly: "It is almost inevitable that human designers are inclined to replicate their own conception of intelligence in the intelligence of robots. In turn, the former gets wired into the control algorithm of the robots" (ibid.: 1511). This leaves users and research about possible demands, pitfalls and forms of usage (including creative use or even adaptation of technology by users; see Oudshoorn & Pinch 2007) out of the picture principally, which in my view is bad advice for any assessment process. Moreover, cultural differences in the perception of robots and interaction with them are only mentioned with respect to the difference of Western and Asian societies. All other differences are flattened by an overgeneralized notion of "the human" – empirical differences between different user groups (constituted e.g. by professional background, affinity to technology, generation) are not taken into account at all. And third, as already indicated, possible

---

[6] One could even argue that any use of a normal PC with internet connection is much more critical concerning most of these issues than at least a single autonomous robot is. For this line of argument see e.g. Brooks 2002.

problems for humans "cohabiting" (that is: interacting) with robots are presented as being the "psychological consequences" of a matter of principle: an ontological confusion which asks too much from ordinary humans. In consequence, empirical research on human-robot-interactivity is not mentioned at all. These three points will structure the course of my argument.

2. Service Robotics – Coordination Practices in a Massively Heterogeneous Field

In the newer approaches to TA mentioned above the basic idea is to try to influence the development process based on the principles for avoiding the possible negative impacts identified, if for the time being. This preferably should be achieved at an early stage of development, to become part of the discussions about design to avoid the fate of many TA recommendations to be doomed to populate the habitat of ELSI-recommendations ("ethical, legal, and social aspects"), which are treated as nice-to-have accessory parts by the developers and which are skipped if it comes to the basic issues, an experience all social scientists had to make more than once.

But coming close enough to the development process and to engineering practice is not easy in general, but especially not in heterogeneous fields. In the case of the "New Robotics" it has to be considered that many robotics researchers indeed agree that robotics has reached the "human frontier" where ordinary humans as the most challenging environment have to be taken into account somehow. But the heading "Human-Centred and Life-Like Robotics", with the last chapter on roboethics, is but one of seven parts of the Handbook (consisting of sixty-four chapters), and most of the specialized questions in the other parts are presented as having nothing to do with it – their content is purely technical. This is not surprising, because a robot is a complex machine by nature – humans in the environment only (but dramatically) enhance this complexity. Robotics researchers are specialized either in kinematics and motion control, research in materials including actuators, all kinds of sensors and sensor analysis, or sensor-fusion, localization, action planning and reasoning, or design of architectures and middleware, to name but a few. "Socially intelligent" or "socially interactive robots" of course enhance the list of necessary disciplinary contributions, as depicted in the following graphics:

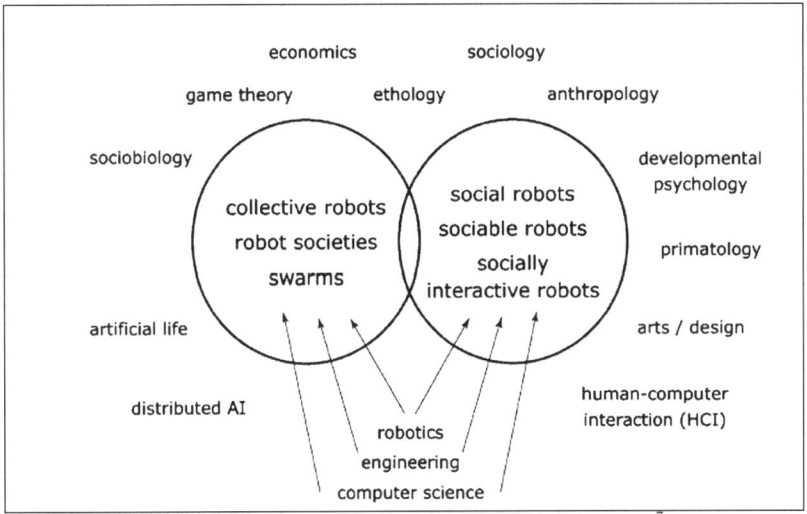

Socially interactive robots according to Fong et. al. 2003: 145.[7]

Furthermore, researchers and developers involved are scholars from disciplinary traditions like mechanical engineering or electrical engineering, different and often competing schools of computer science or AI, materials science, biology, and so forth – traditions that are not only different, but often disregard or even disrespect each other mutually for decades. They often do not understand or accept each others theoretical tradition or even their understanding of "theory", and the families of mathematical calculation they use[8]. And they do not agree at all on application visions, which is the target of many TA-research – for some, especially from computer science and AI, grand visions (like: Computers will beat the human chess champion in five years, or: Robots will beat the human football champion in fifty years) were and are an important driver of their

---

[7] In their overview, the authors distinguish between "collective robots" from "social robots". In the latter class, "socially intelligent robots", for which social interaction plays a key role, form one subclass.
[8] Everyone who ever attended a conference of the "New Robotics" (and not an invited talk by e.g. Rod Brooks) shares the observation of waiting for the consequences of this new philosophy while attending myriads of talks about advances of some mathematical equation types against others, or about purely technical measures - and about approaches presented as unusual where the sociological observer knows that there are classics from he fifties.

field, most of the more engineering oriented researchers believe this addiction to grand visions is a kind of a sickness that does harm to the development of useful machines as to societal debate.

In such a massively heterogeneous field, it seems near to impossible to align with the developers, simply because when it comes to practice they obviously disagree with each other in a more dramatical way than any involvement of social scientists, people from ethics or TA, or ordinary people as users, could ever bring about. In this situation, a discourse (spoken or written words in the natural language) about application visions is not a candidate for TA to become part of the process. Science and Technology Studies (STS) have investigated such heterogeneous fields, and identified some coordination practices within them.

One example is the distinction of classes of algorithms in the field. These – not in detail, but as generalized classes of mathematical reasoning – denote fundamentally distinct approaches to different classes of problems. In robotics, to choose a class of algorithms like neural networks, fuzzy, or genetic algorithms is (even in a very abstracted, often graphically depicted way) understood as a deviation from trying to explicitly model (mathematically describe and formally analyze) every relevant aspect of the application domain, an approach for which the "Good Old AI" stands. This is quite similar to the historical analysis of inter- and transdisciplinary coordinating effect Galison 1997b (see also Duncker 2001) described as the "trading zone" for the early phase of computer simulation, where mathematical "pidgins" allowed mutual understanding and cooperation for a vast range of heterogeneous disciplines.

Another example is using graphical representation as a "tool for collective thinking" (Henderson 1998). In robotics, this is almost contiguous for representing the overall design architecture of a robot, the way its reasoning process ("sense - think - act") and the interplay of it's' components is organized. These graphical representations can be scaled from a hand-drawn sketch, which only grasps the basic relations up to more and more detailed drawings manufactured with CAD-programs.[9]

---

[9] Sociologically this can be described as the coordinative capacity of "boundary objects" (Star & Griesemer 1989), which "are both plastic enough to adopt to local needs and the constraints of the several parties employing them, yet robust enough to maintain a common identity across sites. They are weakly structured in common

It is a major finding of STS-research from the last years that both mathematical and even more graphical representations form the "lingua franca" of engineering sciences and AI. So it is no surprise that basic principles and models in robotics are also communicated in and coordinated via these 'languages'. But this finding becomes a problem if researchers from the humanities and social sciences should enter development fields for conducting TA or roboethics, simply because they are not trained in understanding these languages or even more in using them. So mathematical and graphical representations seem not to be good candidates as entrance point to the development process. But there are at least two other candidates.

In most of the major textbooks of the "New Robotics" a historical line of artificial creatures is drawn, reaching back to classical Cybernetics (e.g. Brooks 2002: 17ff; see also Brooks 1999 and Mitchell et. al. 2000), and these creatures are presented as embodying the grand ideas of Cybernetics: first the possibility of a universal description of humans and machines, which allows to closely couple their activities; second the need for a wide interdisciplinary cooperation not restricted to engineering and math, but including medicine, neuro-psychology and sociology; and third (quoting Norberts Wieners later manifestos) an urge for global responsibility. Though drawing a development line of artificial creatures is but a kind of an unofficial history of robotics agreed upon by some researchers[10], it evocates broader issues that go beyond the "human frontier" as a challenge for machine building only. It is nice to see that this historical reference is also made in the roboethics article of the Handbook:

> "Even a restricted knowledge of cybernetics and computer science, from Wiener, to von Neumann, to Weizenbaum, will immediately and directly demonstrate that these scientists immediately took care

---

use, and become strongly structured in individual-site use " (ibid.: 393). This means that representation of classes of algorithms and robots architectures can at the same time be used very detailed for a specialized public, but also in a very broad and unspecific sense – they can be scaled.

[10] It has to be taken into account that an official history of robotics research does not exist, at least when compared with the importance and the degree of codification of historical reflections in the social sciences and humanities. See for an interpretation of the path that cybernetic universalism took from World War Two to the periphery of the sciences and decade later back on the scientific stage as the historical consciousness of "New Robotics", and of the unofficial and object-centred form of the Cybernetic prehistory of robotics (Meister & Lettkemann 2004).

of the ethical and social aspects of their discoveries and realizations, which marked the beginning of the field of computers and robotics." (Veruggio & Operto 2008: 1522; see also e.g. Murphy 2000: 442)

The reference to cybernetics as a basis for the "design philosophy for service robots" (Kawamura et. al. 1996) comes in two different versions. Most often, the notion of "man-robot-symbiosis" means to model the whole system at large, with humans as only one element in an encompassing system dynamics, thus putting, as it is formulated, "the human in the loop". This version of course is consonant with the hard-boiled (not at least: military) origins of cybernetics[11]. But there is a second version, which concentrates on the history of tiny creatures built to "engage us, entertain us, and enlighten us" (Fong et. al. 2003: 161). For today's more technically advanced situation this version is described as follows:

> "Humans and robots must be able to coordinate their actions so that they interact productively with each other. It is not appropriate (or even necessary) to make the robot as socially competent as possible. Rather, it is more important that the robot be compatible with the human's needs, that it matches application requirements; that it be understandable and believable, and that it provide the interactional support the human expects … Thus, although it is important to continue enhancing autonomous capabilities, we must not neglect improving the human–robot relationship. The challenge is not merely to develop techniques that allow social robots to succeed in limited tasks, but also to find ways that social robots can participate in the full richness of human society." (ibid: 160f)

In this version of the cybernetics reference of the design philosophy, it clearly is the robot that is positioned "in the loop" of human ordinary life, and, as I can add, restricted by it. To summarize this point: The Cybernetics reference can, as it is part of at least a part of the field of the "New Robotics", aid as an entrance point to the inner discussions of the development field.

---

[11] The cybernetic universalism, its historical origin in WW2 and its turn to global responsibility is reconstructed by Galison 1997a, 1998, Bowker 1993 and Hayles 1994. Its prehistory in radar development is reconstructed by Mindell 2001.

3. Investigating Human-Robot Interactivity in Realized Scenarios

As a potential entrance point for TA, in my view the most promising coordination practice in the field is Robocup. Robot tournaments became almost ubiquitous in the last decade, with Robocup as the most prominent event in the entire field with thousands of participants contributing every year. This tremendous success has to be interpreted as the fields answer to the absence of any standardized measures or even benchmarks for a comparison of the performance of advanced robots. The creatures of the "New Robotics" are single-unit R&D-platforms, in which the specific approach of their creators is implemented, and additionally some of the other areas of research that are necessary to built a complete robot – but by nature never all of these areas, because this could only be achieved in a really huge collaborative project encompassing dozens of labs and disciplines. As single-unit platforms inhabiting their creators' labs, these creatures can hardly be compared and overall progress in the field at large can hardly be measured. Moreover, given the heterogeneous nature of the whole field depicted above, the different disciplines involved would even strongly disagree on the 'really important aspects' to become part of a measure – or of a benchmark – for all approaches to "New Robotics".

In this situation, the brilliant trick of Robocup is to combine three organizing principles in one scenario: first, a very simple measure; second, well-known rules; and third a detailed and permanently negotiated description of the setting. Robocup is about winning a series of soccer games where the players are robots. The measure is as simple as it gets – physically scoring more goals that the other teams. This is tremendously demanding task for robots, because a good performance of many technical components is necessary, and all of these components have to work together in real-time. Losing a game can have many reasons, ranging from scientific interesting problems (like error in self-location or object recognition) to more mechanical problems (like running out of battery supply or loosing parts), but can also include human faults by the programmers (like overestimating the robustness of a program in a noisy environment or failed last-minute debugging efforts). These reasons for loosing a game can be very interesting from a scientific standpoint, and researchers can learn a lot from them. But in the game, excuses – other that in a laboratory environment[12] – are ruled out. The ball has to be behind the goal line. The rules of the game itself are also simple; they are imported from the realm of the humans and only slightly adapted to robots.[13]

In contrast to the simplicity and rigidity of the measure and the rules, the setting, consisting of the specification of the playground, the robots and the software platform, is permanently negotiated by all potential participants (mainly in mailing lists). The results are drawn together and fixed by a technical committee, and only in the few days of the tournament itself the rules are fixed. In these negotiations the participants have to find a compromise for two different types of requirements – those of measurable development progress in many research areas, and those of assuring a fair game itself.[14] Besides permanent negotiation of the specification of the setting, this is achieved by a division of different leagues with robots of same size, degree of autonomy, and shape (with humanoid robots as the 'Champions League').

For many researchers Robocup is directly connected to the development of service robots. This sounds a bit strange, because at first sight robots playing soccer with other robots is not what one would expect from robots passing "the human frontier". But in these soccer games, many of the advanced components needed for such advanced robots are tested, including components necessary for interaction. And the developers expect that especially advances in the

---

[12] I have seen many situations in labs, where researchers tried to manually push a robotic research platform out of the door, explaining that the navigation module is at the moment under revision. In Robocup, there is no possibility of this kind of excuse.

[13] Just as in human soccer, there are fixed rules: If a robot looses physical parts it counts as injured and has to be "healed" at the sideline, not able not influence the course of action until the next goal. Just physically overrunning or blocking a robot of the other team is a "foul", and so forth. And there are referees who decide the cases of doubt.

[14] On the one hand, the organizers have to assure that the state of the art of diverse research areas can be addressed in *one* setting. Size and colour of parts of the playing field, for example, must be designed in a way that it is difficult, but doable for the robots recognition program to identify them. This assures that the experiences and the failures during a robot soccer tournament can be published with respect to the state of the art. On the other hand, there are the requirements of the game as a game. Cheating has to be avoided just as any trick of bypassing the competition of complete robots, for example by concentration on physical strength or one per se superior software component.

humanoid leagues will, because humanoids look and behave like humans, foster acceptability of these creatures, without any evidence that this can hold true for real humans in real everyday situations. But the successful drive of Robocup has now reached this class of environment for the robots. Only some years ago Robocup@home started as a new league for everyday tasks in living environment – a living environment of humans. In this league, basically the same organizational principles for the scenario are applied. The setting consists of a few rooms with walls, doors, some furniture and some human volunteers with which the robot has to interact. The measure is the successful execution of tasks that consist of consecutive subtasks, in a given time. An example for such a task is: Enter the home, roll to the living room, identify one of the volunteering persons, ask for her name, navigate to the kitchen, identify a bottle and grasp it, bring it back to the person identified before, hand the bottle to her, and leave the home. The subtasks are tailored to test the performance of different technical components, but for the whole task all of the components have to work together in real-time. There is not much specification of the robots themselves, because size and technical functionalities follow from the setting. Only external communication with the human team members is explicitly forbidden. As in the soccer case, only success counts. There is a scoring system in which every successfully executed subtask within the time given sores a predefined number of points. Again, there are no excuses for observable failure, e.g. for not finding the entrance to the test area at the very beginning – this happens, and then the score is zero. And again, explicit part of the rules is benevolent behaviour of the human teams. That means, "trying to cheat (e.g. pretending autonomous behavior where there is none)" and "trying to exploit the rules (e.g. not trying to solve the task but trying to score)" (RoboCup@Home Technical Committee 2009: 16) is not legitimate. And as in the soccer case, the specifications are negotiated by potential participants.

So Robocup@home applies the principles for Robocup to the realm of service robots, making single-unit platforms from the labs comparable. But given the more complex mission, it is not surprising that not everything can be imported from soccer tournaments. The measure lacks the undisputable evidence of the soccer case, so the tasks and their rules are also part of negotiations, and to my observation the referees have a much more important role. Avoiding too much standardization is, at least to some extend, even intended by the organizers to address the ever-changing character of a realistic real-world environment:

> "To foster advance in technology and to keep the competition interesting, the scenario and the tests will steadily increase in complexity. While in the beginning necessary abilities are being tested, tests will focus more and more on real applications with a rising level of uncertainty." (ibid.: 5)

But the most important point here is that human-robot interaction is for the time being not very developed in the setting. Despite aiming at "natural human-robot interaction" (ibid.), the volunteering persons in all tasks are strictly instructed to speak loud and clearly, not to move at all, and avoid anything that could irritate the robot. These are not very "natural" circumstances.

Given the explicit aim to foster developments for better human-robot interaction, it is intriguing that for the design of the tasks and the setting no link at all to the respective research community is made. This is "I-Methodology" again. To be sure, the brilliant trick of Robocup is simplification (only counting observable results in a realized scenario[15]), but one would expect that the makers of RoboCup@Home would derive at least some challenging settings from empirical investigations made by the human-computer interaction (HCI) or the human-robot interaction (HRI) research communities. And a closer look at these empirical investigations can inform the further development of robot tournaments. All of these investigations show, that in real interaction with a robot, and with respect to the humans' acceptance of robots, different groups of humans feel and behave in very different ways. There are different proposals for basic distinctions in groups: From differences in generation (Sackmann & Weymann 1994) to more abstract types of users (Rogers 2003) or a typology of roles humans can take towards a robot (supervisor, operator, mechanic, bystander and teammate role; Scholtz 2003). Some of these basic distinctions have to be chosen, if

---

[15] For the case of a non-robotics area (Ubiquitous Computing) we have discussed some of the advantages of applying physically realized scenarios for the observation of human-computer (and sensor etc.) interaction as opposed to just asking abstractly about it with e.g. questionnaires (Meister et. al. 2008).

the question how classes of ordinary people[16] will react to advanced robot behaviour in their everyday environment. Just positioning volunteers from the participating teams will of course never address this issue. Another important issue from HRI-research is a fundamental scepticism about assumptions like "a nice look" or human-like behaviour of the robots per se guarantees smoother interaction with "the human".

> "The criteria for 'good performance' often differ substantially. In particular, 'functionally designed' social robots may need only to produce certain experiences for the user, rather than having to withstand deep scrutiny for "life-like" capabilities." (Steinfeld et. al. 2006: 38)

Because it lies in the logic of the further development of RoboCup@Home to include more realistic human behaviour in the setting and in the tasks, one can assume that a (not to complicated) differentiation of "humans" in different groups will somehow appear and will somehow affect the specification of the setting of the tournament, as well as the question of acceptable shape and behaviour of different robots.

But giving the robot developers the advice to include empirical findings of HRI-research in their considerations – very practical, as hints for the further advancement of the rules and the setting of RoboCup@Home – faces the problem that HRI-research is heterogeneous in its own. Approaches range from statistical measurements of human factors to quantitative model of interaction up to more qualitative models and 'softer' ethnographic observations of human-robot interaction "in the wild". Researchers involved in the rather new field of HRI perceive the lack of any possibility to compare the results of case studies and the lack of common measures for good or smooth interaction as the major weakness of their field.

In the field, synoptic approaches, which order all relevant aspects and results in a unitary way – or in a repository – are seen as the solution. For example, Steinfeld et. al. 2006 present a "HRI metric toolkit and

---

[16] Just one example is reported in Häussling 2009 where students as probands were more sympathetic to the robot if they were from the social sciences, while students from computer science were more suspicious because they believed that they where cheated by a "wizard of oz" (a programmer bypassing the technical functionality 'by hand'), an assumption that derives from their professional experience.

reference source", and Weiss et. al. 2009 present an encompassing "evaluation framework for Human-Robot Collaboration addressing usability, social acceptance, user experience, and societal impact", complemented by an assignment of appropriate methods and a multi-level indicator model. From the TA perspective it is noticeable that many of the issues from roboethics appear in these repositories as lists of empirical indicators for "human effectiveness" or "social impact". But such an ordering of aspects is not selective and somehow conservative, which seems to be the reason why the use of expert evaluation and user studies (including scenarios) is restricted to aspects of classical usability, while social acceptance, user experience and societal impact are only assigned to questionnaires and focus groups.

An alternative approach is to include some if not all of these issues in a basic model of human-robot interactivity and empirically investigate them in realized scenarios or in field studies. Interactivity stands, other than the much broader term interaction, for an investigation of the processes between two different entities (or species) where the result emerges from the process itself. This approach has requirements for the conceptualization of humans and robots. From a social science perspective this means that the "action is not only with the humans", but agency is "distributed between humans, machines, and programs" (Rammert 2008). From the HRI perspective this means to take into account that at least advanced robots do not have complete knowledge of their environment, but "intelligent behaviour is ... the ability to retrieve information through interaction" with humans (Weiss et. al. 2010: 38). So in his approach modelling has three interconnected parts: the human, the robot, and the process between them (see e.g. Häussling 2009 as an example), where the results of this process can be learning and adaptation of expectations and habits – on both sides, literally for humans, in its technical counterparts for the robots.

4. Conclusions

Roboethics explicit goal is to foster mutual discussion, and this is also a goal of newer approaches in TA. I tried to sketch some points where the great divide between the two versions of the "human frontier" – humans as a challenge and robots as a thread – could be bridged by relating the practical approaches in a heterogeneous development field and the results of different strands of research like Robocup@home and HRI. Instead of drawing on general assumptions about "the human" (again: from both sides) the empirical investigation of human-robot interactivity in realized scenarios can reveal what groups of ordinary people find acceptable, useful or even attractive in robots. Moreover, adaptation of technology to human habits and creative usage of artefacts can be identified. Thus, from a TA perspective, realized scenarios are one of the "loci" to investigate and hopefully influence "promises-requirement-cycles" in the technology development process (Rip & Shot 2002).

This of course does not mean that there are no critical issues about the robots entering our daily environments, which call for guidelines within the research process and especially for applications. And of course not every technical tool or gadget that users find acceptable, useful or even attractive makes sense from the perspective of society at large. But my feeling is that the concentration of roboethics on principal psychological problems, which originate from an ontological confusion between the human and the artificial, is inadequate – this question should be left to the assessment of ordinary humans. And for robotics development I think that instead of the grand "human frontier" visions the more modest approach that derives from Cybernetics history is preferable in trying to achieve ways in which humans and robots could be "able to coordinate their actions so that they interact productively with each other" (again: Fong et. al. 2003: 160). In other words, the possibilities of mutual augmentation of humans and robots in interactivity should be investigated.

References

Akrich, M. (1995): "User Representations: Practices, Methods and Sociology", in: Rip, A. Misa, T. J. & Schot, J. (eds.): *Managing Technology in Society: The Approch of Constructive Technology Assessment*. Pinter Publishers: Londen/ New York, pp. 167–184.

Billard, A.; Calinon, S.; Dillmann, R. & Schaal, S. (2008): "Robot Programming by Demonstration", in: Siciliano, B. & Khatib, O. (eds.): *Handbook of Robotics*, Springer, Berlin, pp. 1371–1394.

Bowker, G. (1993): "How to be universal: some cybernetic strategies, 1943 – 1970", *Social Studies of Science*, Vol. 23, pp. 107–127.

Breazeal, C.; Takanishi, A. & Kobayashi, T. (2008): "Social Robots that Interact with People", in: Siciliano B. & Khatib, O. (eds.): *Handbook of Robotics*, Springer, Berlin, pp. 1349–1370.

Brooks, R. A. (1999): *Cambrian intelligence: The early history of the new AI*, MIT Press, Cambridge, Mass.

Brooks, R. A. (2002): *Flesh and machines: robots and people*, Pantheon Books, New York.

Duncker, E. (2001): "Symbolic Communication in Multidisciplinary Cooperations", *Science, Technology, & Human Values*, Vol. 26, pp. 349–386.

Fong, T.; Nourbakhsh, I. & Dautenhahn, K. (2003): "A survey of socially interactive robots", *Robotics and Autonomous Systems*, Vol. 42, pp. 143–166.

Galison, P. (1997a): "Die Ontologie des Feindes. Norbert Wiener und die Vision der Kybernetik", in: Rheinberger, H.J.; Hagner, M. & Wahrig-Schmidt, B. (eds.): *Räume des Wissens. Repräsentation, Codierung, Spur*, Akademie Verlag, Berlin, pp. 281–324.

Galison, P. (1997b): *Image and logic. A material culture of microphysics*. University of Chicago Press, Chicago, Il.

Galison, P. (1998): "The Americanization of unity", *Daedalus*, Vol. 127, No. 1, pp. 45–71.

Guston, D. H. & Sarewitz, D. (2002): "Real-time technology assessment", *Technology in Society*, Vol. 24, pp. 93–109.

Häussling, R. (2009): "Video Analysis with a four-level Interaction Concept: A network-based Concept of Human-Rotot Interaction", in: Kissmann, U.T. (eds.): *Video Interactiom Alalysis. Methods ans Methodoloty*, Peter Lang, Frankfurt a.M., pp. 107–131.

Hayles, K. (1994): "Boundary disputes: Homeostasis, reflexivity, and the foundations of cybernetics", *Configurations*, Vol. 3, pp. 441–467.

Henderson, K. (1998): *On line and on paper. Visual representations, visual culture and computer graphics in design engineering*. MIT Press, Cambridge, MA.

Kawamura, K.; Pack, T.; Bishay, M. & Iskarous, M. (1996): "Design philosophy for service robots", *Robotics and Autonomous Systems*, Vol. 18, pp. 109–116.

Meister, M. & Lettkemann, E. (2004): "Vom Flugabwehrgeschütz zum niedlichen Roboter. Zum Wandel des Kooperation stiftenden Universalismus der Kybernetik", in: Strübing, J.; Schulz-Schaeffer,

I.; Meister, M. & Gläser, J. (eds.): *Kooperation im Niemandsland. Neue Perspektiven auf Zusammenarbeit in Wissenschaft und Technik*, Leske + Budrich, Opladen, pp. 105–136.

Meister, M.; Pias, M.; Töpfer, E. & Coulouris, G. (2008): "Application Scenarios for Cooperation Objects and their Social, Legal and Ethical Challenges", in: Michahelles, F. (ed.): *First International Conference on The Internet of Things. Adjunct Prodeedings*, Zürich, pp. 92–97.

Mindell, D. A. (2001): "Automation's finest hour: Radar and system integration in World War II.", in: Hughes, T. P. & Hughes, A. C. (eds.): *Systems, experts, and computers. The systems approach in management and engineering, World War II and after*, MIT Press: Cambridge, Mass., pp. 27–56.

Mitchell, R. J.; Bischop, J. M.; Keating, D. A. & Dautenhahn, K. (2000): "Cybernetic approaches to artificial life", *Künstliche Intelligenz*, Vol. 1, pp. 5–11.

Murphy, R. R. (2000): *An introduction to AI robotics*, MIT Press: Cambridge, MA.

Oudshoorn, N. & Pinch, T. (2007): "User-Technology Relationships: Some Recent Developments", in: Hackett, E.; Amsterdamska, O.; Lynch, M. & Wajcman, J. (eds.): *The Handbook of Science and Technology Studies*, MIT Press, Boston/ Mass., pp. 541–567.

Prassler, E. & Kosuge, K. (2008): "Domestic Robotics", in: Siciliano, B. & Khatib, O. (eds.): *Handbook of Robotics*, Springer, Berlin, pp. 1253-1282.

Rammert, W. (2008): "Where the Action is: Distributed Agency between Humans, Machines, and Programs", in: Seifert, U.; Kim, J. H. & Moore, A. (eds.): *Paradoxes of Interactivity. Perspectives for Media Theory, Human-Computer Interaction, and Artistic Investigations*, transcript, Bielefeld, pp. 62–91.

Rip, A. & Shot, J. W. (2002): "Identifying loci for influencing the dynamics of technological development", in: Sørensen, K. H. & Williams, R. (eds.): *Shaping technology, guiding policy: Concepts, spaces and tools*, Edward Elgar, Cheltenham, pp. 155–172.

RoboCup@Home Technical Committee (2009): *Rules & Regulations. Final Version.* http://www.ai.rug.nl/robocupathome/documents/rulebook2009_FINAL.pdf (29.06. 2009).

Rogers, E. M. (2003): *Diffusion of Innovations*, Free Press, New York u.a.

Sackmann, R. & Weymann, A. (1994): *Die Technisierung des Alltags. Generationen und technische Innovationen*, Campus, Frankfurt a.M.

Scholtz, J. (2003): "Theory and evaluation of human robot interactions", *Proeedings of the Hawaii International Conference on System Science* (36).

Schot, J. & Rip, A. (1997): "The Past and Future of Constructive Technology Assessment", *Technological Forecasting and Social Change*, Vol. 54, No. 2-3, pp. 251–268.

Siciliano, B. & Khatib, O. (eds.) (2008): *Handbook of Robotics*. Springer, Berlin.

Star, S. L. & Griesemer, J. R. (1989): "Institutional ecology, 'translations' and boundary objects: Amateurs and professionals in Berkley's Museum of Vertebrate Zoology, 1907–39", *Social Studies of Science*, Vol. 19, pp. 387–420.

Steinfeld, A.; Fong, T.; Kaber, D.; Lewis, M.; Scholtz, J.; Schultz, A. & Goodrich, M. (2006): "Common metrics for human-robot interaction", in: *First ACM International Conference on Human Robot Interaction, Salt Lake City, UT*, pp. 33 – 40. Available at: http://www.ri.cmu.edu/pub_files/pub4/steinfeld_aaron_m_2006_1/steinfeld_aaron_m_2006_1.pdf, (12 July 2006).

Veruggio, G. & Operto, F. (2008): "Roboethics: Social and Ethical Implications of Robotics", in: Siciliano, B. & Khatib, O. (eds.), *Handbook of Robotics*. Springer: Berlin, pp. 1499-1524.

Weiss, A.; Bernhaupt, R.; Lankes, M. & Tscheligi, M. (2009): "The USUS Evaluation Framework for Human-Robot Interaction", *AISB2009: Proceedings of the Symposium on New Frontiers in Human-Robot Interaction. SSAISB*, pp. 158-165. http://www.aisb.org.uk/convention/aisb09/Proceedings/NEWFRONTIERS/FILES/WeissABernhauptR.pdf, (12 July 2009).

Weiss, A.; Igelsböck, J.; Tscheligi, M.; Bauer, A.; Kühnlenz, K.; Wollherr, K. & Buss, M. (2010): "Robots Asking for Directions - The Willingness of Passers-by to Support Robots", in: *5th ACM/IEEE International Conference on Human-Robot Interaction, Osaka*, pp. 23–30. https://www.lsr.ei.tum.de/fileadmin/publications/HRI2010final.pdf (25 July 2010).

Technology Assessment of Service Robotics. Preliminary Thoughts Guided by Case Studies

Michael Decker

Abstract: Recent developments in service robotics prove that service robotics is undergoing the same success story that industrial robotics experienced before. While service robots are still closely linked to professional human workers for surveillance and agricultural purposes, non-professionals such as senior citizens or patients might use service robots for their health care. These other application areas demand a high grade of adaptability and learning capabilities of the technical systems. In this paper this so called technology push scenario of service robotics is described and transformed into a first set of research questions from the perspective of technology assessment.

Keywords: Technology assessment, service robotics, autonomous systems, learning systems

1. Introduction

Following a classical distinction in innovations research[1] different perspectives can initiate technology assessments (TA): the "technology push" and the "demand pull" perspective. On the one hand, new technologies are constantly being developed whose intended and unintended consequences should be investigated within the social context of their application ("technology push" perspective – technology-induced TA). On the other hand, social needs or problems can be identified which can possibly be solved or their consequences moderated by using technical processes ("demand pull" perspective – problem-induced TA). Even if this distinction initially creates an artificial division – ultimately it is of course the aim of a comprehensive TA to judge the opportunities and risks of technical procedures in the light of the social context of their application and thus overcome this division[2].

At the beginning of technology assessment, the "technology push" arguments came to the fore. In the discussion prior to the institutionalization of the Office for Technology Assessment (OTA) at

---

[1] A sketch of the history of "technology push" and "demand pull", including critical statements, can be found in Nemet 2009.
[2] This has also been proposed in the context of innovations research (see, for example Nelson & Winter 1977).

the US Congress, policy makers especially needed advice on technical developments, such as experimental aircrafts, anti-ballistic rockets, and large technical plants (Grunwald 2002: 101). Hence, the Technology Assessment Act of 1972 reads as follows: "To establish an Office for Technology Assessment for the Congress as an aid in the identification and consideration of existing and probable impacts of technological application" (US Senate 1972). This general demand of policy makers was confirmed in connection with other parliamentary TA institutes. The Commissions of Inquiry that were dealing with the preparation of the Office for Technology Assessment of the German Bundestag (Federal Parliament) are quoted (after Paschen, Petermann 2005: 15) as follows:

> "The German Bundestag has to fulfill significant tasks in the field of science and technology. This does not only include the evaluation and control of various complex and expensive research and development projects and programs of the government, but also – in the light of the growing depth of engagement of science and technology into nature and society and the widespread skepticism towards the implications of new technologies – the design of the framework conditions of technological change and the participation in the societal dialogue on chances and risks of science and technology. The German Bundestag is only very insufficiently prepared for these tasks, it does not have enough possibilities – independent from the executive – to obtain, process, and assess the required information. Technology Assessment could be a suitable instrument for that and appropriate action should be taken to make TA available for the German Bundestag." [Translation by the author] (Paschen, Petermann 2005: 15)

Grand global challenges like climate change, sustainable water management, or nuclear waste disposal were not the first to put societal needs in the focus of TA, too. Technical applications are being developed to contribute to the fulfillment of these needs. TA then evaluates these technical solutions in comparison with other, also non-technical, alternatives. The impacts of technical developments and processes are in the focus of attention of TA, which is emphasized in the German term "Technikfolgenabschätzung" (assessment of technological impacts). In the methodological reflections these impacts or consequences are identified to be "unintended", "undesired", "unexpected", etc. (Gloede 2007).

Nowadays, TA defines itself as a problem-oriented approach[3], which is reflected in the latest definitions: "Technology Assessment is defined as scientific and communicative contributions to resolve technology-related societal problems"[4]. A group of TA practitioners from leading European TA institutions, who have jointly worked out the following definition for the EU project "Technology Assessment in Europe: Between Method and Impact (TAMI)" phrased it in a similar way: "Technology Assessment is a scientific, interactive and communicative process which aims to contribute to the formation of public and political opinion on societal aspects of science and technology."[5]

This definition underlines the fact that TA can contribute to the public and political opinion, but does not make the related decisions on its own. TA generates knowledge that contributes to the solution of societal and political problems related to technology, but is neither able nor legitimized to solve these problems on its own.

The relevant criteria proposed for the problem analysis are the same, irrespective if the TA practitioner got involved in the topic from the "technology push" or "demand pull" perspective. The criteria developed in the TAMI project should be mentioned here as an example (Bütschi et al. 2004):

- Issue dimension
  The first – and trivial – dimension is the issue to which a problem is related. When assessing the situation, it is important to be aware of the way the issue is framed. It might be a *technology-oriented* issue, i.e. the problem is connected to a concrete technology, such as transportation systems, waste-treating technologies, energy supply techniques or concrete treatments in biomedicine. Since technology intervenes in many domains of our life, such as health, work, entertainment and mobility, it might be sensible to frame a problem as a *domain-oriented* issue. Examples are e-commerce or e-health. Finally, the *consequence-oriented* approach might be the most adequate. In this context, the project does not mainly address the technology, but puts the emphasis on societal trends or changes that are technology-related. Typical examples are projects

---

[3] Bechmann & Frederichs 1996.
[4] Grunwald 2002: 52.
[5] Bütschi et al. 2004: 14.

addressing the questions of privacy, sustainable development, gender division, or North/South relationship.
- Political dimension

  Issues addressed by TA are generally politically relevant. This relevance, however, might change depending on the applicable stage of the policy-making process and on the nature of the ongoing political debate. In the *agenda-setting* phase, government has not yet officially addressed an issue. In the *policy-making* phase, an issue is already on the political agenda and at the stage where fundamental decisions have to be taken. The *policy-implementation* phase represents the stage of the policy cycle in which a clear policy on the issue at stake (e.g. fostering e-learning) has been formulated, but the policy still has to be implemented. Finally, the political problem might be *political deadlock,* i.e. no solution is in sight. Examples are the European debate on genetically modified food and the debates on nuclear waste disposal sites in many countries.
- Social dimension

  Another characteristic aspect is the social dimension of the issue under consideration. The *value dimension* is inherent to every technology, as has been shown by TA in the past few decades. However, there are differences in how far-reaching the relevance of values is in a concrete case and situation. The *relation to the public* is of crucial relevance. Is the issue already creating public interest? How is the interest expressed if it exists? Is there fascination, rejection, mistrust? Who is leading the social discussion? Are large organisations (parties, churches, social movements) aware of the issue? *Social roles and relationships* must be considered. The design of a TA study may crucially depend on the assessment of the roles of experts, decision-makers and laypersons and their mutual relationships in the respective field.
- Innovation dimension

  Along the development path of a specific technology, TA has different entry points. TA-relevant questions alter in the various phases of the innovation cycle, and there are different stakeholders and social groups involved. Accordingly, fitting TA questions to the development phase of the respective technology and to the corresponding decision-making requirements is an essential element of assessing a situation. Following the widely used model

of the innovation chain, we were able to identify the following different stages of development: early stages of research and development, industrial research, the marketplace, widespread diffusion, and embedded technological systems.
- Availability of knowledge
  TA must provide and manage knowledge. Knowledge generation in TA has its own specific difficulties because anticipatory and therefore hypothetical knowledge is required. The design of a TA study depends on the amount and quality of knowledge already available in the respective situation and on identified knowledge deficits and gaps. Therefore, an exploration of the availability of knowledge belongs to each pre-phase of TA. Different points of departure in this field are: high-quality knowledge available; high degree of consensus among experts and scientists; high-quality knowledge available only in some relevant fields of the issue under consideration, with other areas of ignorance or high uncertainty; or knowledge is available about gaps in knowledge ("*gewusstes Nichtwissen*" "acknowledged ignorance").

This paper takes a "technology push" perspective on service robotics in order to think about a technology assessment by looking at service robotics as it is described in the pertinent science and technology journals, at specialist congresses, in reports on the work of expert committees, in patents, etc. These can be understood as arenas in which the "enactors" of new technologies present and promote their concepts, from which the "selectors" of new technologies then draw their information, and in which both groups conduct a debate (VanLente & Rip 1998, Rip & te Kulve 2008). The basis of this screening is thus a comprehensive search of documents for statements about new technological developments and the justification for their relevance given by the enactors, which is corroborated particularly by the expectations expressed in terms of social and political trends, developments, needs, and problems.

2. Robots and Service Robots – Definitions

In general, there is a distinction drawn between two types of robots: service robots and industrial robots. Industrial robots are established in almost all areas of the manufacturing industry. The automotive industry, just like metalworking, plastics, rubber, timber, and furniture industry, is barely conceivable without industrial robots.[6] Over the last few years the world market for industrial robots has grown continuously, however, not in all regions of the world to the same extent (IFR 2008). Features of industrial robotics include high speed and precision, enormous power and an almost unlimited repeatability of movements in combination with little downtime. From the economic point of view human output has been replaced by technological output or, to put it more simply: Labor costs have been replaced by costs of acquisition and operation. The fact that the complete production process had to be rebuilt for the use of robots is not a technical problem at all. A production hall is a confined space and its interior is optimized for the production process and designed according to the regulations for safe production and occupational safety.

Almost the same innovation potential as for industrial robots has been predicted for "service robots" for some time now. First of all, the term service robots seems to cover all "non-production robots" (see below). A closer look at the areas of application of service robot systems sold worldwide until the end of 2008 reveals that out of 63,000 service robots for commercial applications the greater number were used in the field of defense, rescue and safety (30%), followed by agriculture (23%), especially milk robots (International Federation of Robotics (IFR) 2009). These are areas where the service robots are operated and supervised by a human expert and/or in a protected surrounding. This can therefore be interpreted as a transition zone between industrial robotics and general service robotics. The robot itself is no longer active within its "safety cage" which is normally set up for a safe production process. However, outside its cage it is only used in areas where it generally does not come into contact with a third party or does not carry out services aroud human beings. The person who cooperates with the robot can be trained for this cooperation, which turns him – to a certain degree – into a robotics expert himself.

Most services, however, are characterized by the fact that they have to be performed in an environment full of people (one example might be the cleaning of train stations) or directly involve a human being

---

[6] According to the Verband Deutscher Maschinen- und Anlagenbau e.V. (German Engineering Federation), in 2006 already 50% of all industrial robots were installed in other sectors than the automotive industry.

(museum guide, nursing or elderly care). The people in contact with these robots can only be trained to a limited extent as robotics experts. Thus these services implicate that a robotic layperson can and has to interact with robots and that third parties will encounter a robot's direct environment. Furthermore, these services are performed in everyday life, which can only be adapted to a limited extent to the employment of robots.

First thoughts about a working definition of service robots against this background include the fact that was already mentioned above: service robots are "non-industrial robots" (conf. tab 1, List of Service robots by IFR). IFR states on its website:

> "Service robots have no strict internationally accepted definition, which, among other things, delimits them from other types of equipment, in particular the manipulating industrial robot. IFR, however, have adopted a preliminary definition:
> A service robot is a robot which operates semi- or fully autonomously to perform services useful to the well-being of humans and equipment, excluding manufacturing operations."[7]

We will first draw our attention to robots in general. A closer look at the history and development of robots, described – among others – by D. Ichbiah, reveals that robots set a milestone in the progressive human attempt to create machines which enable people to perform better, take over some of their workload, support them and finally are subservient and undemanding servants[8]. The term robot harkens back to the Czech author Karel Capek. In his native language "robota" means servant or obedient worker. A technical definition can be found in the VDI guideline 2860 (Assembly and handling; handling functions, handling units, terminology, definitions, symbols):

> "A robot is a freely and repeatedly programmable, multifunctional manipulator with at least three independent axes to move material, parts, tools or special instruments on programmed, variable tracks to fulfill various tasks." [Translation by the author] (Christaller et al. 2001: 18)

The following definition adds the fact that the robot is "supporting the human being" to the technical description:

---

[7] IFR: http://www.ifr.org/service-robots/ last viewed on 20.05.2010.
[8] Ichbiah 2005: 9ff and in particular 26ff.

> "Robots are sensorimotor machines to extent the human ability to act. They consist of mechatronic components, sensors and computer-based control functions. Robots are extremely complex; more degrees of freedom as well as the variety and extent of their forms of behavior distinguish them considerably from other machines." [Translation by the author] (Christaller et al. 2001: 18)

This definition attempts to differentiate between robots and simple automats by pointing out the larger number of degrees of freedom and the variety and extent of its forms of behavior. An automated garage door or a bread maker feature mechatronic components, sensors and a control function, but they are not complex enough according to the above definition. A modern aircraft or automobile is also equipped with the mentioned technical elements and is much more complex. According to the above definition, they could be classified as robots. Something similar is true for ubiquitous computing. Whole apartments that are equipped with the technical features of Ambient Assissted Living[9] could be described as robots corresponding to the definition above, even if they are normally not included in this subordinate concept.

Turning to the service robots now, we first of all should define if the term "service" is used according to everyday language usage or in an economic science context[10]. "Service" in the colloquial sense can be described as "the sum of all human work steps [...] that satisfy needs directly without the use of material goods".[11] So the focus is on executing a service, accomplishing or acting in general instead of material goods.[12] From the economic point of view the term "service" seems to be insufficiently defined. This might be due to the extremely multifaceted types and forms of services and the fact that they cannot be clearly distinguished from contributions in kind or material goods. According to Maleri, "services" are intangible assets produced for a

---

[9] For examples see http://www.aal-deutschland.de/.
[10] The European Union law also gives a short definition of services in the context of "freedom to provide services" in the treaties of the European Union: Services shall be considered "services" within the meaning of the Treaties where they are provided for remuneration, in so far as they are not governed by the provisions relating to freedom of movement for goods or capital, like activities of commercial character or craftsmen.
[11] Maleri 1997: 6.
[12] ibid.

third-party need using external production factors. At the same time "production" is defined as the directed fabrication of material goods and services using other material and immaterial goods and is divided into different (economic) sectors:

- Primary industry (primary production) which covers agriculture, forestry, fishing, hunt and sometimes also mining;
- secondary industry (secondary production) with the manufacturing industry and craft as well as
- the service industry (tertiary production) which includes all other economic sectors like commerce, banks, insurances, restaurants, consulting etc.

In addition, a further division of the tertiary sector into a quaternary or quinary sector for information and leisure is being discussed[13]. According to Clark a distinction can also be made between direct and indirect services. While the end user is the direct user of direct services, indirect services are production factors. Intangible real goods can be subdivided into performance, services, information and rights. Especially the distinction between performance and services seems to be relevant in the context of robotics and highlights again the fuzziness of the everyday usage of the term "service". Performance is understood as the physical and mental human performance provided by households. Although it is also a characteristic of numerous services, in the end it is an isolated, non-complex offer, i.e. not a good resulting from the use or the combination of several production factors (auxiliaries, supplies, current assets, planning, organisation, ...).[14]

The definition of service robots reveals the reference to the economic concept of services. In 1994, the Fraunhofer Institute for Manufacturing Engineering and Automation (Fraunhofer IPA) phrased the following definition of the work of the Institute, which is still valid today:

> "A service robot is a freely programmable mobile device carrying out services either partially or fully automatically. Services are activities that do not contribute to the direct industrial manufacture of goods, but to the performance of services for humans and

---

[13] Maleri 1997: 10f.
[14] ibid.: 23 and 53.

institutions." [Translation by the author] (Schraft, R. D. et al. 2004: 9)

The following definition highlights the specific characteristic that distinguishes service robotics from industrial robotics:

> "Robots in the service sector will differ from industrial robots; they will be individually designed for the execution of a given task taking place in a specific environment."[15]

Before we conclude our considerations on the definitions of service robotics with a tabular classification (table 1), we will quote a definition of Engelhardt and Edwards after Hüttenrauch:

> "[…] systems that function as smart, programmable tools, that can sense, think, and act to benefit or enable humans or extend/enhance human productivity." (Hüttenrauch 2006: 3)

The word "think" in this definition explicitly points out cognitive skills. This definition of service robots can be combined with definitions of robots in general which refer less than the ones mentioned above to the technical equipment as central defining element. Trevelyan (1999), who refers to "intelligence", can be quoted as an example: "Robots are intelligent machines to extend human skills". This definition brings up other terms like "intelligence", "autonomy", "cooperation", etc. which have to be discussed in the context of interdisciplinary technology assessment.

---

[15] Schraft et al. 1993.

Technology Assessment of Service Robotics

Personal/Domestic Robots	*Professional Service Robots*
Robots for domestic tasks • Robot butler/companion/assistants/humanoid • Vacuuming, floor cleaning • Lawn mowing • Pool cleaning • Window cleaning Entertainment robots • Toy/hobby robots • Robot rides • Pool cleaning • Education and training Handicap assistance • Robotized wheelchairs • Personal rehabilitation • Other assistance functions Personal transportation (AGV for persons) Home security & surveillance	Field robotics • Agriculture/Milking robots • Forestry • Mining systems • Space robots Professional cleaning • Floor cleaning • Window and wall cleaning (including wall climbing robots) • Tank, tube and pipe cleaning • Hull cleaning (aircraft, vehicles, etc.) Inspection and maintenance systems • Facilities, Plants • Tank, tubes and pipes and sewer • other inspection and maintenance systems Construction and demolition • Nuclear demolition & dismantling • Other demolition systems • Construction support and maintenance/Construction Logistic systems • Courier/Mail systems • Factory logistics (incl. Automated Guided Vehicles for factories) • Cargo handling, outdoor logistics/Other logistics Medical robotics • Diagnostic systems • Robot assisted surgery or therapy • Rehabilitation systems • Other medical robots Defense, rescue & security applications • Demining robots • Fire and bomb fighting robots • Surveillance/security robots • Unmanned aerial vehicles/Unmanned ground based vehicles Underwater systems Mobile Platforms in general use Robot arms in general use Public relation robots • Hotel and restaurant robots • Mobile guidance, information robots • Robots in marketing • Others (i.e. library robots) Special Purpose • Refueling robots • Others Customized robots Humanoids

Tab. 1: IFR: List of service robots by tasks

3. Case Studies

The "problem field" service robotics shall be examined on the basis of four case studies. Since different cases are expedient for technology assessment, the differences will be roughly analyzed according to the TAMI criteria in a second step. First of all, the case studies shall be described in the sense of a "technology push". This means that we will refer to the descriptions of technology developers and – if already available – also quote first "reflections" from the societal perspective.

1.) Robots in agriculture

At the conference on "Automation and Robots in Agriculture" of the Association for Technology and Structures in Agriculture Arno Ruckelshausen stated that the development of autonomous field robots marks the next step of the inevitable automation of agricultural technology. In the years after the introduction of the first marketable prototypes for specific tasks like weed and pest control and the respective information on the economic, ecological or social framework conditions a coexistence of large agricultural machines and small field robots has to be expected.[16]

All this started more than 20 years ago in greenhouses with harvesting robots for tomatoes and cucumber. Soon afterwards the robots also found their way into horticulture as harvesting robots for citrus fruits and apples. Already in 2006 Baerveldt and Astrand of Halmstad University introduced the first weedkilling robot.[17] To date, other developers also try to establish such weedkilling robots as fully functional autonomous machines on the fields. Projects like BoniRob, an autonomous field robot to collect measured data of individual plants developed by the University of Applied Sciences Osnabrück in cooperation with Bosch and the Amazonen-Werke, which shall later be used for targeted weedkilling or fertilizing of individual plants, or "weed killer", a robot developed by the École Nationale d'Ingénieur de Brest and AGRO DEAL for weedkilling in row crops can be mentioned here.[18]

---

[16] Ruckelshausen 2010.
[17] Grift 2007.
[18] Ruckelshausen/AMAZONE: „BoniRob-Feldroboter schafft Basis für die Landtechnik der Zukunft" at: www.info.amazone.de/DisplayInfo.aspx?id=13763; sowie Chocron et al. 2007.

Characteristic features of these field robots are the ability to move freely, i.e. autonomously on a defined area and the special emphasis put on imaging techniques which enable the robots to differentiate between the moisture content of plant and fruit or identify shapes and sizes or seed rows. The aim is to specify and improve the efficiency of pest control and fertilization of agricultural crop to increase the harvest using a constant amount of resources and preserving the environment at the same time.

In this context, the term pest control is not limited to herbizides. SlugBot is a robot developed at the University of the West of England to eliminate slugs, which are not only by horticulturalists and agriculturist regarded as vermin and are therefore combated.[19] SlugBot is moving autonomously across the field using DGPS data. In doing so, it illuminates the soil with red light to detect slugs, i.e. longish, flat shapes on the ground, which the camera displays as white spots on black background. If a slug is detected, SlugBot collects it with its grappler. Its maximum speed is 10 slugs per minute. Another particularity of this robot is the subsequent use of the slugs: The collected slugs are transported to a separate off-board digester unit. They are converted into methane and used to generate electric energy for the robot itself.[20]

SlugBot is being described as energy efficient and precise in executing its job. At least for professional agriculture it therefore seems to be a real alternative to other methods to combat harmful slugs, which often have a number of disadvantages.

Apart from the motives described so far – pest control and micro fertilization – there is one other important reason for the use of field robots that has to be mentioned: The acquisition of information on the status of field crop and soil before and during the growing season. So-called scouting robots, possibly in the form of swarm robots, shall be used to collect the required information on soil condition, soil fertility and soil contamination, growth and condition of the crop, moisture content or infestation by pests and make them available to the farmer or trigger and coordinate the use of more specialized robots.[21]

Another categorie of robots in agriculture are automatic steering aids. Fields are normally cultivated in parallel rows (adjusted to the terrain).

---

[19] Schraft et al. 2004: 70.
[20] Schraft et al. 2004: 70.
[21] Grift 2007.

The difficulty for the farmers is that there should be neither gaps between these tracks nor should they overlap. Steering systems installed in their agricultural machinery shall enable them to work more precisely. The systems vary in their complexity but they all rely on GPS signals to determine their position during parallel tracking: The basic models consist of lightbar displays for the driver's platform where flashing lights indicate the direction in which the driver has to steer. But first of all a reference track is needed and the track width has to be determined. The so-called steering assistants are more complex and rather recognized as robots. They can be retrofitted and interfere directly with the steering (in most of the cases at the steering wheel) and guide the tractor automatically down the tracks. These systems consist of three units: The receiver for the GPS signal that every system needs and which is normally installed on the roof of the vehicle; an operating and computer unit with display to control the functions of the steering system, for documentation and data management including screen mapping, for ISOBUS accessory control as well as for an overview on the performance features like efficiency, fuel consumption, area, time, total output, etc.; and the steering unit which is responsible for directional control. Higher accuracy can be achieved with permanently integrated steering systems which have a shorter delay from signal input to steering movement. Setting up base stations to calculate the position coordinates with Real Time Kinematics (RTK) also improves the precision to centimeter-level accuracy.[22][23]

Although the farmer is still sitting on the driver's platform, he no longer has to navigate thanks to the automated steering; he is rather controlling and managing. Finally the aim of using these autopilots is to cut operating costs. The more precise field work without overlaps and skips reduces the number of runs on the field and thus helps to save fuel and time. Seed, fertilizers and pesticides can be placed more precisely and without waste; especially row crops can be planted more efficiently and the use of machines also facilitates harvesting. GPS systems make it possible to work on the field in low visibility situations caused by dust or at night. This makes the farmer more flexible. The physical strain for the machine operators is also reduced

---

[22] John Deere 2009: Brochure on Guidance Systems.
[23] Heckert & Lenge 2007.

and they are completely relieved from the monotonous but concentration-demanding task of lane-keeping.[24]

Steering systems are paving the way for so-called electronic towing bar systems which allow the control of other autonomous agricultural vehicles within one working process (e.g. combine harvesting, dispatching the harvested produce and baling) by wireless modems from a single-manned tractor. According to Patrick O. Noack, this towing bar allows a single worker in the field to control vehicles with a much higher combined performance and traction.[25]

2.) Driver assistance systems in motor vehicles

Driver assistance systems in motor vehicles can be roughly divided into two categories: "on board" systems, which do not receive external information (from external systems) and telematics systems which aim at a superordinate coordination of the whole traffic. These are called "Adaptive Cruise Control" (ACC), the 2nd generation of cruise control systems. A logic unit calculates the distance to and the speed of the preceding vehicle using the information transmitted by the sensors and interferes – if necessary – with the engine control and the brake system. Usually this sensor is a radar device. Often one radar device is used for the distant areas more than 30 metres away and an additional one or a camera system for the close-up range. If the safety distance to the car in front is less than the preset value, a three-tier anticipatory emergency braking system is activated. The first one is the collision warning, which alerts the driver by brake jerks, warning signals and pretensioning the seat belt. This is followed by the emergency braking assist, which preventively prepares the brake system for an emergency brake by pre-charging the brakes, placing the brake pads against the discs and reducing the activation threshold of the hydraulic brake system. If the driver does not react sufficiently to prevent a collision with the vehicle ahead, the third stage, the automatic emergency brake is being activated. It acts on all relevant systems, including a full braking, to avoid a crash. However, just like the previous steps it is first of all meant to assist the driver and shall not replace him. Once the driver touches the gas pedal the ACC system automatically

---

[24] See ibid.
[25] Noack 2010.

accelerates the vehicle again. Thus the system can also be perfectly used in stop-and-go traffic to support the driver.[26]

According to V. Gruhn the term "vehicle telematics" is a collective term for the compilation, transmission, processing, and use of traffic-related data aiming at the organization, information, and control of traffic to pursue the following goals:

- Increase the mobility
- Increase the utilization of the different means of transport as well as the capacity and efficiency of the traffic infrastructure
- Charge user fees
- Increase road safety
- Reduce environmental pollution (CO_2, particulate matter, etc.)[27]

Telematic traffic systems like Traffic Message Channel (TMC, also called Radio Data System (RDS); since 1997 nationwide available), the satellite-based freeflow toll system Toll Collect (since 2003/2005 in Germany) or traffic control systems like variable message signs have long found their way onto German roads.[28]

Current efforts of car and traffic development aim at anticipatory driver assistance systems. The focus is here on digital navigation maps that communicate with the Advanced Driver Assistance System (ADAS) as well as on Vehicle-to-Infrastructure (V2I) and/or Vehicle-to-Vehicle (V2V) communication ((extended) Floating Car Data). The common goal of theses technologies is to extend the driver's anticipation beyond his own field of vision and that of the sensor systems (camera, radar) installed in the car by early and exact information. This shall result in increasing the safety by allowing for more response time and thus avoiding obstruction of traffic.[29]

"Floating Car Data (FCD) "exploits" vehicles to collect traffic data. These vehicles "float" with the traffic and can therefore be used to

---

[26] BOSCH: Adaptive Cruise Control, unter: www.bosch-kraftfahrzeugtechnik.de/ media/de/pdf/fahrsicherheitssysteme_2/adaptivec-ruise controlsicherheitsabstandeingebaut.pdf. Stand: April 2010.

[27] Gruhn, V.: Telematik III: Verkehrstelematik. Institut für Informatik, Universität Leipzig. – part 1.

[28] BMVBW 2004; sowie Toll Collect: FAQ: http://www.toll-collect.de As at April 2010.

[29] Nöcker et al. 2005.

determine the average cruising speed. Especially equipped vehicles, digital image processing, registration at cell sites or toll booths can be used for this purpose. Extended Floating Car Data (XFCD) uses terminals in the car to get access to data of the vehicle electronics. Information from ABS and ESP can be used to locate icy conditions on the road."[30] [Translation by the author] This information will be sent directly or indirectly to the following road users, or to be more precise, their car systems which process it by calculating an alternative route, adapt the speed or preset the electronic safety systems.[31]

Theoretically a direct regulation of assistance systems (like cruise control) by extended traffic control systems would also be possible. Instead of just showing a variable traffic sign, a vehicle with cruise control could be set to the adapted maximum speed. The traffic flow could be improved.

3.) Robots in private life (children, young adults, elderly, people in need of care)

Service robots in private households are the subject of the third case study. Robotics provides applications for all age groups: Toy robots, entertainment robots, kitchen aids, assistant robots, care robots for elder and sick people, etc. The private domain is hardly monitored. A private user of a robot has only few duties like reading the user manual, complying with service intervals etc. This filed of application puts high demands on the robots. They have to be able to move in an unknown environment (apartment) that is not geared to them and perform a number of different tasks. If the programming efforts prior to the start of operation should be still acceptable for robotic laypersons, most of the adaptation to the new environment and the new user has to be done by the robot itself. This technical problem is even more critical for older users and those in need of care, since they are often cognitively unable to give the necessary instructions.

Thus the "intuitive" handling of the robot system is becoming more important and therefore, at least according to some supporters of humanoid robot systems, they should look as human as possible to lower the levels of acceptance. This can be a decisive factor if the

---

[30] Gruhn, V.: Telematik III: Verkehrstelematik. Institut für Informatik, Universität Leipzig. – part 2.
[31] Nöcker et al. 2005.

robot has to replace a person providing a service, especially in the field of care.

The private domain is closely linked to the privacy of the individual. On the one hand, the use of service robots can protect the privacy of the individual against insights of other people – which might be desirable for instance in the genital area – but on the other hand the robot and its technical conditions might make the privacy of the individual even more visible and controllable for third parties.

A broad variety of applications of robots for children is discussed: Toy, artificial pets, learning aid, babysitter, substitute teacher, etc. The term "edutainment" (Schraft et al. 2004) combines elements from two areas which are intended for completely different purposes: "Education" in the form of training, teaching, and learning is often the exact opposite of "entertainment". To categorize them as toy robots (Ichbiah 2005, Wehmeyer et al. 2010) is also difficult since this includes a number of very different robot systems. Therefore some robot systems will be briefly described here as examples:

"Pleo" is a robot in the shape of a dinosaur baby (Camarasaurus) developed by Innvo Laps. Being equipped with two microphones, two loudspeakers, a camera, and approx. 14 sensors at head, chin, shoulders, back and legs under the skin, Pleo can get in contact with the outside world.

What is special about Pleo is that he can interact with his environment and show his feelings by certain facial expressions, gestures, or sounds. Pleo "is hungry", "wants to play", "needs attention", "is tired". Interaction with Pleo, like stroking his head or back, touching his legs, simulating to feed him, playing with him, talking to or ignoring him, satisfies his needs and thus he develops, learns, is being "educated".

A number of robotic building sets were developed in the tradition of LEGO® or fischertechnik building sets (ROBO by fischertechnik and MINDSTORMS 2.0 by LEGO®)[32]. The focus of these products is on programmable logic components that are equipped with numerous connections for sensors and actuators as well as interfaces to a computer. Pathfinders, sorting systems or other exciting installations can be easily assembled. With the appropriate directions and guidance, these building sets will give children from the age of 8 the

---

[32] http://www.fischertechnik.de/de/index.aspx and
http://www.technik-lpe.eu/produkte/lego-education/lego-mindstorms.html.

opportunity to gain their first experience in the field of robotics. By now a number of challenges have been established around schools and universities, e.g. FIRST® LEGO® League, where the teams have to complete various tasks with their self-made robots. There are even special activities for girls like ROBERTA[33] to awaken the girls' interest in (this) technology.

PaPeRo (Partner-type Personal Robot), a robot developed in Japan, can be described as a further development of babyphones – some kind of babysitting robot. "Our continuing research & development is geared toward creating communication robot that can live with us and serve as companion to all of us including children and the elderlies."[34] PaPeRo features voice recognition, voice response, facial recognition, face tracking and touch sensors. It is mobile, recharges its batteries automatically, can imitate sounds and play a quiz. In addition, this robot can be used to send messages. It can be controlled remotely and its software can be enhanced by open access. PaPeRo is intended as a companion for children[35].

Different types of household robots have already been developed, above all those, which are already "in use" like vacuum cleaning or lawn mowing robots. The kitchen seems to be an area where robot assistance is especially welcomed. In contrast to the vacuum cleaning robot, which *replaces* the human being as operator of the vacuum cleaner, here the focus is on the cooperation with the human being. The humanoid robot ARMAR, which was developed by the Collaborative Research Center "Humanoid Robots - Learning and Cooperating Multimodal Robots" (SFB 588), shall be employed in the kitchen. The aim of the project is to develop concepts, methods and concrete mechatronical components for a humanoid robot, which will be able to share its activity space with a human partner. In order to be a helpful assistant in everyday life, the robots system must have many complex abilities and characteristics: Armar 3 is for example able to bring small items like cups, mugs, a pack of rice, or a juice box. It can also fetch a particular drink from the fridge, lay the table, or load and

---

[33] http://www.iais.fraunhofer.de/uploads/media/Roberta_Mappe.pdf.
[34] http://www.nec.co.jp/products/robot/en/index.html. As at June 2010.
[35] This idea is not further explained on the website of the manufacturer. However, numerous blog entries and on-line magazines discuss PaPeRo and other robots of its kind, like e.g. Rogun by KornTech, as the new generation of babysitters.

unload the dishwasher. Learning by demonstration is a central element of the cooperation between human and robot in SFB 588.

The scientific competition "Robocup@Home" also put the application area "at home" into a scientific focus. It is about household service robot; the in-field testing takes place in realistic apartments (living room, kitchen or even garden). The robots are completely autonomous and equipped with "intuitive" human-machine-interfaces like natural language, gestures, etc. The following topics are of interest for the robotic researchers at Robocup@Home (see http://www.robocup.org/robocup-home/):

- Human-Robot-Interaction and Cooperation,
- Navigation and Mapping in dynamic environments,
- Computer Vision and Object Recognition under natural light conditions,
- Object Manipulation,
- Adaptive Behaviors,
- Behavior Integration,
- Ambient Intelligence,
- Standardization and System Integration.

Submissions addressing practical applications using international benchmarks and/or in-field experimental testing are strongly encouraged.

Different service robots for the support of elder or sick people are either in development or already in prototype status. Since these robots have already been described in numerous other publications and are a recurring topic of public discussion we would like to refer to the relevant literature.[36]

4.) Robots as mediators

Taking up the ideas of these three case studies we have a fourth one which examines robot systems as mediators between humans and between humans and technology. The research field "humanoid mediators" studies and develops anthroposophical systems according to Prof. Tanja Schutz (KIT), which allow for human-environment-interaction and interpersonal communication across all boundaries and restrictions. The ways of human communication are limited by spatial

---

[36] For example: Weber 2006 or Decker 2008.

or temporal distances and cultural and linguistic communication difficulties. Physical or mental restraints and impediments limit the human capabilities to interact with the environment. To guarantee a realistic interaction and communication across such obstacles forms, topics and modalities of human actions and intentions have to be recognized, interpreted, and transferred in a way that preserves the meaning and makes them experienceable with all senses. This act of *mediation* shall be provided by anthropomatic systems reliably and efficiently at any time. Humanoid systems shall be used as ideal mediators here. The term humanoid systems covers two different systems: humanoid robots which fit into the human environment thanks to their humanoid appearance and encourage an intuitive natural interaction, but also systems directly linked to the human body like substitutive and additive prostheses which can expand the abilities of a human due to precise tailoring to the individual. In addition, humanoid systems are seamlessly integrated into the network of internet, devices, systems and technologies and have thus always comprehensive access to all digital information. These humanoid mediators shall act as a link between the world of humans and machines. Considerable advantages are attributed to humanoid mediators compared with human middlemen: They are always available, never get tired, are replaceable, neutral in terms of content, impartial and ubiquitous. Humanoid mediators shall understand and interpret human actions, expectations and intentions to communicate them realistically and preserving the meaning. Receivers of these transmissions are both the digital and the analog environment of the human being as well as the human interaction partners, which are linked *by* the humanoid mediator across all barriers. Humanoid mediators are relevant for people with one or several of the above mentioned impediments or impairments who are unable to interact or communicate with their environment or the people around them and are therefore not or only in part able to participate in social life. In addition, humanoid mediators could be of interest for institutions that operate locally or globally but depend on a reasonable and efficient communication of their employees, the possibility to train them from the distance with real objects and integrate the scattered personnel tightly into the work processes.

4. Discussion

The case studies outline the technical features that are associated with service robots from the "technology push" perspective. In the following we will now add some first thoughts on the overall societal situation around these technologies to this technical dimension. This will be done with reference to the "dimensions" developed in TAMI.

1.) Service robots in agriculture

Agricultural service robots are not on the agenda of political discussion, although agriculture itself and the respective grants are a regular issue in political debates. Agricultural robots are also not in the focus of society. There are discussion forums for individual applications (e.g. milk robots). Agriculture is known as a producing industry mainly characterized by (large) machinery. From an economic point of view, which also considers its innovation potential, agriculture will be assigned to primary production according to the above-mentioned categories. The robots are operated in a professional environment inside and outside the "production halls". The most common agricultural robot systems, the milk robots, require the adaptation of the environment (i.e. the cowshed) to the robot – just like in industrial robotics. Also other robot systems (like autonomously driving tractors) are being operated under professional surveillance. Farmers who operate robot systems can be requested to attend relevant trainings etc. In 2008, approx. 15 000 systems were sold world-wide, so this is a sector where robots are already used. Therefore it is also possible to gain empirical insights: forums like "Indoor agriculture, Buildings and Plants" on the website landlive.de provide discussions on "Milk robots – yes or no?". Regarding their degree of innovation, there are already different forms of milk robots on the market. Other service robots, like self-driving tractors, are currently still in prototype state.

2.) Driver assistance systems

Driver assistance systems can help to achieve various aims in transportation policy. The establishment and further development of driver assistance and telematics systems (including, among others, Adaptive Crusie Control (ACC) to contribute to active road safety or Vehicle-to-Infrastructure Systems (V2I) to allow a communication (data transfer) between vehicle and infrastructure to control the traffic) are based on the following reasons:

A.) Road Safety: Technical systems shall improve the active road safety.

B.) Optimization of traffic flow: Technical systems shall optimize both the economic and ecological aspects of the motorized individual traffic. Road traffic shall become more efficient and congestions shall be avoided.

This will be emphasized by the following statement by the Federal Ministry responsible for transport issues:

> "Within its research program, the Federal Ministry of Transport, Building, and Housing (BMVBW) will analyze strategies, methods and technologies for the improvement of traffic safety and traffic flow as well as the management and control of traffic to improve the efficiency of the existing infrastructure capacity."[37] [Translation by the author]
>
> "Even more than to date, vehicle technology shall be used to avoid accidents (active safety) and to minimize the consequences of accidents, i.e. to improve the passive safety. The use of telematics systems in road traffic will contribute to the avoidance of accidents." [Translation by the author] (ibid.: 17f)

Regarding road accidents it can be stated that the number of people killed in road traffic is steadily declining. This is on the one hand due to measures like the introduction and lowering of the legal blood alcohol concentration limit or the introduction of helmet laws and seat belt legislation, but on the other hand also a result of the vehicles' increased passive safety (e.g. airbag, ABS).[38] Not only has the number of people killed in road traffic been reduced, but also the overall number of accidents. While there were 413,942 accidents with damage to persons due to driver error registered in 2005 in Germany, this figure had declined to 388,201 in 2008. Taking a closer look at the causes of the accidents it can be noticed that 15% were attributed to ignoring the right of way, 14.35% to inappropriate speed and 11.5% to not respecting the safety distance. So 40% of all road accidents with damage to persons result from these three failure causes.[39] An analysis

---

[37] BMVBW 2001: 8.
[38] Statistisches Bundesamt 2006: 48f.
[39] ADAC: Verkehrsunfälle in Deutschland. Statistiken http://www1.adac.de/ Verkehr/Statistiken/default.asp?id=430&location=2_Verkehr. As at April 2010.

of rear-end collisions with injured people made by BOSCH revealed that 20% of the drivers applied the brakes too late, 50% did not brake hard enough and 30% did not brake at all.[40]

The assumption is that current and future technology could be used to automatically keep the safety distance to the vehicle in front and adapt the speed to the traffic situation and traffic rules, i.e. to decelerate the vehicle and thus possibly reduce the number of collisions.

Traffic congestions are an everyday phenomenon. In the German state of North Rhine-Westphalia there are at least 100 congestions per day.[41] Congestions often occur at crossroads and traffic lights, in bad weather, because of roadworks or accidents. According to M. Schreckenberger, every German citizen is caught up in congestion for an average of 2.4 days per year. The cost for a three-hour congestion with a length of 4 km on a two-way "autobahn" (German freeway) amount to € 20,000 to 100,000.[42]

Dietmar Bachmann, member of the State Parliament of Baden-Württemberg, reported during the 32. session of the Parliament on 11 September 2007 that "the cost for congestions that occur in road traffic due to the waste of fuel [...] [amount] to approximately 12 billion € per year. The total economic loss due to congestions on our roads amounts to more than 100 billion € per year."[43] [Translation by the author] Already back in 2001, the then State Secretary at the Federal Ministry of Education and Research Uwe Thomas stated that an additional 12 billion liters of fuel were consumed in traffic congestions, which equals to an economic loss of 190 billion DM.[44]

Apart from the aforementioned causes, we can refer to two other reasons for congestions, which are related to driver assistance systems: "Phantom traffic jams" and route suggestions of GPS navigators.

The so-called "phantom traffic jam" forms at places with a high density of traffic – for whatever reason – typically on expressways. A traffic volume of more than 1500 to 1800 vehicles per lane and hours

---

[40] BOSCH: Notbremssystem http://www.bosch-kraftfahrzeugtechnik.de/media/ de/pdf/fahrsicherheitssysteme _2/ vorausschauendesnotbremssystem _hilftauffahrunfllezuvermeidenmindertunfallfolgen.pdf. As at April 2010.
[41] NRW: 2009.
[42] Weltonline: Bundesbürger stehen 535.000 Jahre im Stau, 13.09.2009.
[43] Landtag BaWü 2007.
[44] http://www.innovations-report.de/html/berichte/verkehr_logistik/bericht-4197.html. As at April 2010.

makes the traffic instable. If a vehicle comes in such a situation to a halt due to inattention or overreaction of the dirver, it takes him approx. 2 seconds to accelerate. This time span is longer than the time span to approaching vehicles which are therefore also forced to brake – thus a small disturbance leads to a traffic wave that travels upstream with approx. 15 km/h.[45] According to Schreckenberg, the underlying problem for the occurrence of congestions in heavy traffic is that people do not respect the safety distance to the vehicle ahead while they also drive at inappropriate speed. In Germany, the statutory safety distance corresponds to a time interval of 1.8 sec, independent of the speed. The reaction time is slightly above 1 second.[46]

ACC Systems could help to keep the safety distance to the vehicle in front and optimize the acceleration after braking. "Controlled driving" with adjusted speed and the necessary safety distance should allow for a traffic flow of up to 2200 vehicles per line and hour without congestions.

Navigation systems can be combined with TMC receivers (Traffic Message Channel). These devices use the TMC traffic reporting to warn of danger or to automatically calculate individual routes according to the current traffic situation to avoid congested areas. TMC can only transmit information on expressways and major highways, but hardly on rural or urban roads. At the same time these GPS navigators tempt the drivers to leave the congested roads. However, highways, rural or district roads are not designed to accommodate the traffic from the expressway.[47] ADAC, the German Automobile Association, meanwhile advises the drivers to "stand" the congestion on the autobahn instead of following the alternative routes on smaller roads calculated by the GPS navigator.[48] In this context telematics systems can be used to control the "overall system" on all types of roads.

This overall situation illustrates that transportation is not only a permanent topic in political discussions but also a central element in

---

[45] cf. Schreckenberg 2007.
[46] ibid.
[47] Weltonline: Bundesbürger stehen 535.000 Jahre im Stau, 13.09.2009; sowie ADAC: Staufrei mit TMC Aktuelle ADAC-Verbraucherberatung. http://www1.adac.de/images/Staufrei-mit-TMC-Aktuelle-ADAC-Verbraucherberatung-2004_tcm8-89477.pdf. Stand: April 2010.
[48] ADAC: http://www1.adac.de/Auto_Motorrad/Technik_Zubehoer/Navigation/Probleme/default.asp#atcm:8-211285 As at April 2010.

social perception. Driver assistance systems mark a further step of robots "leaving the factory floors". The system of individual traffic can be described as a grown infrastructure with various established rules. Every vehicle has got an owner. He has several obligations like reading the manual, ensuring the roadworthiness of the vehicle, regular general inspections, third-party insurance, driver's license, etc. The road traffic regulations provide the legal framework for this part of society. In general "everybody" should be able to operate a vehicle – if no health restrictions apply. This means vice versa that a training as "robotics expert" is hardly possible for every vehicle driver. In addition, the vehicles are operated in public, where "third parties" – pedestrians, bicyclists, etc. – can also be encountered. There are no driver assistance systems designated for them.

3.) Robots in private life
Robots in private life are not yet on the political agenda. The robot systems that are already established on the market – primarily lawn mowing robots and vacuum cleaning robots – are not really relevant for political discussion. In connection with politically relevant topics like the situation of nursing care in Germany, the feasible technical development of care robots is in the center of political perception. However, marketable robot systems that could be able to take over assistant tasks in households are still in prototype state. The use of robots in sick-nursing and elderly care is also in the focus of public debate. Every now and then other applications in the private sector are being discussed as well, like the question of "living with robots" (Reichert 2009).
There is also another area where the robot can be related to a societal need. In February 2010 it was reported that South Korea is going to introduce robots as assistant teachers in nurseries and preschools in 2012. The program called R-Learning is scheduled to start with 400 robots and will be using 8,000 artificial teachers in 2013. These "teacher robots" shall read stories to the children or serve as monitoring interface for worried parents.[49] Just like computers have found their way as a medium or aid into everyday school life, it can be assumed that robots will also be applied in schools. As already said, South Korea is also pioneering here with the little home robot IROBI.

---

[49] http://de.engadget.com/2010/02/23/r-learning-sudkorea-will-ab-2012-roboter-lehrer-einsetzen/.

Its characteristics are similar to the communication robot PaPeRo and it additionally features, among others, an English tutorial. It can be used as teaching or learning medium. Jeonghye Han et al. (2005) conducted a small survey investigating the effects on concentration, interest and achievement when using the home robot IROBI to learn English compared with other media like book plus audio media or web-based instructions. To this end, pupils of equal capability of the same grade were divided into 3 groups, 5 girls and 5 boys each, to test the three types of media (home robot, web-based instructions, and book plus audio media). They were given 40 minutes to gather the learning objectives of an English lesson. Since this is not a long-term study, the results are still relative, but a first trend can be observed: Concentration, interest and achievement of the experimental group that used the home robot IROBI were higher than those of the other groups using the other media (cf. fig. 2).[50]

TABLE V Average score of instructional effects for each group

Type of media / Factor	Book with an audio tape	WBI	Home Robot	F	p-value
Concentration	2.32 (0.59)	2.85 (0.52)	3.76 (0.21)	23.754	.0000***
Interest	3.3 (0.69)	3.4 (0.82)	4.5 (0.84)	7.041	.0035**
Achievement	3.1 (0.73)	3.3 (0.48)	4 (0.66)	5.482	.0100*

Note: (standard deviation), *** denotes p-value=0.001, ** denotes p-value=0.01, * denotes p-value=0.05

Fig. 2: Home robot IROBI (left)/Results of the study (right)[51]

4.) Mediators

In this series of case studies, mediators are the ones that are furthest away from technical implementation according to their degree of innovation. This is especially true for mediation between different cultures and telepresence applications. Therefore, mediation by humanoid roboters is neither part of the political nor the societal discussion. If the developments of pervasive and ubiquitous computing are included in the considerations, the need for mediation with the technical environment can be anticipated for the near future. One example of medical applications shall be mentioned here as well: The use of robots in the treatment of autistic children. Up to 1.16% of people suffer from autism. These are considerably more people than

---

[50] Han et al. 2005.
[51] ibid.

those who are, for example, affected by trisomy 21 (between 0.125% and 0.2%). The spectrum of autism is very diverse, there are different severities from infantile autism to the Asperger's syndrome, some forms can hardly be differentiated from AD(H)D. In general, autism is classified as incurable. Being described as perception or information processing disorder of the brain or as congenital, abnormal mode of information processing, autism is characterized by impairment of social interactions and communication as well as by stereotypical behaviour patterns. Very seldom strengths in perception, attention, memory and intelligence lead to the savant syndrome. The impairment of the social behaviour results from the peoples' difficulties in talking with others, making or keeping eye contact, and understanding language, facial expressions and gestures, interpreting them correctly or using them appropriately themselves in the respective situations.[52]
Kerstin Dautenhahn et al. developed the idea to use robots from the human robot interaction (HRI) research as mediators in communication processes to help autistic children therapeutically to get into contact with other people. As described in their publications, the idea seems to be successfully further developed and applied.[53] The robot KASPAR is at the heart of this research and therapy measure. KASPAR is a humanoid robot torso, which has the size of a sitting child. It can move its arms and head and has a face to show some simple expressions. Due to the way its shape, movements and especially its face are designed it resembles a human being, but does not appear completely realistic. The movements and facial expressions of KASPAR can be remotely controlled by the patient himself; only if the patient is very young or critically ill the carer will take over the control.

---

[52] Wikipedia: Using the relevant keywords. As at June 2010.
[53] Robins et al. 2009.

Technology Assessment of Service Robotics 81

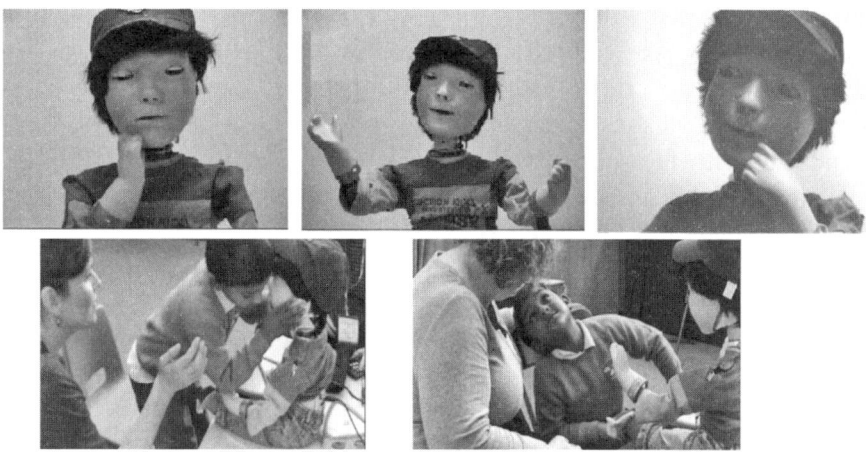

Fig. 3: KASPAR [54]

In all three cases described in the publication at hand by Dautenhahn et al. the autistic children were very interested in KASPAR and felt the desire to touch it and explore its face although they normally did not try to get in contact with other childern or adults (fig. 3). After the experience of touching KASPAR those children started to feel their own faces and also shared their feelings with the people present, which is extremely remarkable. This shows that these autistic children are able to transfer their experience gained from the interaction with KASPAR to other people present. So they were finally able to perceive, look at and touch other people. These are the first steps to improve the communication skills of these children and support them therapeutically. It is assumed that KASPAR works as mediator because of its clinical, object-like but nevertheless humanoid appearance."[55]

5. Conclusion and Outlook

The case studies presented in this article compare industrial robotics and service robotics by trying to illustrate the different levels of application from industrial robot to service robot. In general it can be stated that the agricultural applications can still be described as "industrial robotics". Talking about a milk production industry, a

---

[54] cf. ibid.
[55] cf. ibid.

cowshed equipped with a milk robot can be described as a production site. The distinctive feature is that here cows are being "processed" in large scale production instead of things like cars. Third parties normally do not have access to the cowshed. Those people who have to interact with the robot can be instructed in usual regular trainings. They are trained as some sort of "robotic experts for milk robots", they might even have to take a relevant exam.

Assistance systems change the individual scope of action of each car driver, if the robot system used in the vehicle does not communicate with its environment. The service can then be described as an optimization of the usability. The car can be easier braked, parked, accelerated, etc. However, in combination with a telematics system, the performance of the car in the overall system of traffic infrastructure will also be optimized. We have to analyze in detail according to which criteria the optimization is being carried out here. In this context, the criteria or preferences of the individual (to get from A to B as fast as possible) might conflict with the preferred criteria on the cumulated level of the overall system (to minimize the sum of the journey times of all individuals in the system). This would for example be the case if one individual is advised to make a wide detour around a congested urban area in order to relieve the area from traffic. The robot system is permanently installed in the vehicle and thus becomes a part of the relation between vehicle producer and vehicle owner. Its functions are checked at the regular inspections and its mode of operation is explained to the owner in the instruction manual. However, the information provided therein does not turn the owner into a "robotic expert". But nevertheless the use of a vehicle requires the driver to familiarize himself with the functions of the vehicle prior to the departure.

The use of robot systems in the private sector has different framework conditions. Of course private users have to read and follow the manuals of technical equipment as well. But the knowledge about this (robot) device resulting from this "instruction" is very limited. Especially if the robot is intended for the use with children, elderly or sick people, the robot system has to fulfill different technical requirements regarding safety and handling. In addition, it has to be anticipated that third parties who have no knowledge about the robot systems will encounter the robot in private life. Here the so-called "intuitive handling" of the robot would become relevant, which would also enable third parties to use or at least pass it safely.

Example number 4, the mediator robots, represents a cross-cutting scenario. Here the communication between human and machine is in the focus. This is one prerequisite that has to be fulfilled in the other case studies as well to realize a successful cooperation. However, the concrete technical form of the interface for "communication" is completely different in each case study. While would be rather talking of a control stand to monitor the functions of the milk robot in the agricultural sector, a driver "communicates" with a vehicle in the well-established way of using the pedals, levers, buttons and displays, etc. A brake assistant is activated through the brake pedal without further possibilities of interaction by the driver. If required, a robot system like traction control can be completely switched off by the driver. The requirements on a robot system in the private sector also differ considerably. If the users are able to operate them, keyboards similar to those of personal computers could be applied. Apart from that, systems using natural language are of special interest here. Mediator robots represent an example that can be used to analyze the possibilities and limits of the key element of communication between human and machine. The aspect of communication between people – which is "mediated" by a robot system and, if necessary, in a second step also automatically translated into another language or even another cultural context – represents a particular challenge to the robot systems.

All in all, the four case studies are sufficiently different to serve as a basis for the discussion of the various questions regarding service robotics. They can be differentiated regarding the environment in which the robots can be used:

a.) The (semi) professional environment in agriculture and mediation, where professional translation can also be a topic.
b.) The public space of traffic infrastructure for driver assistance systems.
c.) The private area with service robots that are used as "helpers of all sorts", especially for children and people in need of care.

The case studies also provide a starting point to put the services into larger contexts. The agricultural context is linked with an entrepreneurial activity, which can also be true for the use of driver assistance systems. However, these are also intended for private users of vehicles. In the private sector we can also find applications that can be assigned to business ventures like a leasing model for care robots, while other applications represent private consumer goods. The same applies for the mediator robots.

Finally we will analyze the case studies regarding their relevance for the different scientific disciplines of an interdisciplinary technology assessment. This means that we have to find out whether they can be used as an example to illustrate the issues identified during the formulation of the problem. First of all this can be stated for the technological perspective. Each case study is characterized by different requirements on the robot systems – not only regarding their technical embedding in other (non-)technical systems but also relating to the communication and cooperation with humans. These examples also cover different levels of the innovation process: Systems already on the market, systems in prototype state or close to introduction on the market as well as systems still in the research state are discussed. From an economic point of view both businesses and private households are included on the microeconomic level, but also macroeconomic effects in transportation or health economics are important. The legal perspective is similar. The question of responsibility under private law has to be discussed, for example if the robot does any damage. First of all the existing legal framework has to be adapted to the new warranty and risk problems. This refers to issues of the contractual terms, especially regarding the assignment of the liability risk in the General Terms and Conditions as well as basic questions of liability for damages to third parties. From the perspective of public law, the relation between the state and the citizen is put into the center of attention. The regulation of both the transport and the health sector could be analyzed on this level to find out whether the introduction of robot systems would necessitate corresponding changes. From the psychological point of view it has to be examined how the communication and cooperation between human and robot can be ideally designed in the different fields of application. The focus is here on the question "How humanoid should robot systems be designed?". This does not apply at all for the agricultural sector and can also be neglected for driver assistance systems since the interface is permanently installed in the automobile and therefore no connection with humanoid robotics can be expected. However, the private sector and also mediation are areas where humanoid robots play a major role. In this context it is also important to find out which characteristics people consider as relevant to attribute humanoid features to a robot system and which role non-visual aspects, like natural language, play. On the one hand, ethical questions arise from the philosophical point of view when it comes to determining the societal domains in which a modern society should employ robot systems. Taking the example of elder care or nursing, it is still unusual today to assign certain tasks to technical systems. Therefore a detailed consideration of the different fields of application according to ethical criteria is highly recommended. In connection with the driver assistance systems it also has to be discussed if there is a danger of instrumentalization for the human driver within the whole system. This could be the case if the vehicle either does not accept certain "orders" by the driver because of the current constellation of other system parameters or if superordinate coordination aspects of the telematics system have negative consequences for the individual which results in questions of acceptability. On the other hand, as already expressed, the humanoid appearance of service robots also touches on anthroposophical questions like "Will technical and non-technical ideas of man be changed by humanoid robots?".

Taking all this into account, we can assume that the case studies described here do not only sufficiently cover the field of service robotics with its multifaceted applications but also provide the necessary links for an interdisciplinary consideration of the question. From the methodological point of view it is desirable that – if not all then at least several – scientific disciplines should identify relevant questions on the basis of these case studies. This would further the integration of interdisciplinary knowledges. At the same time it was possible to describe the overall societal and political situation in the sense of a "technology push" problem analysis on the basis of the case studies. In this respect, the presented case studies provide a good basis for an interdisciplinary technology assessment. However, this ex ante evaluation has to be confirmed in the interdisciplinary project phase.

References

Bechmann, G. & Frederichs, G. (1996): "Problemorientierte Forschung: Zwischen Politik und Wissenschaft", in: Bechmann G. (ed): *Praxisfelder der Technikfolgenforschung. Konzepte, Methoden, Optionen*, Campus, Frankfurt/M., pp. 11–37.

BMVBW (2001): *Programm für mehr Sicherheit im Straßenverkehr*, Berlin.

BMVBW (2004): *Telematik im Verkehr. Entwicklungen und Erfolge in Deutschland*, Berlin.

Bütschi, D.; Carius, R.; Decker, M.; Gram, S.; Grunwald, A.; Machleidt, P.; Steyaert, S. & van Est, R. (2004): "The Practice of TA. Science, Interaction, and Communication", in: Decker, M. & Ladikas, M. (eds): *Bridges between Science, Society and Policy. Technology Assessment – Methods and Impact*, Springer, Berlin, pp. 13–55.

Christaller, T.; Decker, M.; Gilsbach, J.-M.; Hirzinger, G.; Lauterbach, K.; Schweighofer, E.; Schweitzer, G. & Sturma, D. (2001): *Robotik. Perspektiven für menschliches Handeln in der zukünftigen Gesellschaft*, Springer Berlin, Heidelberg.

Chocron, O.; Delaleau, E. & Fleureau, J.-L. (2007): *Flatness-Based Control of a Mechatronic Weed Killer Autonomous Robot*, IEEE, pp. 2214-2219.

Decker, M. (2008): "Caregiving robots and ethical reflection: the perspective of interdisciplinary technology assessment", *AI & Society* 22,3, pp. 315–330.

Gloede, F. (2007): "Unfolgsame Folgen. Begründungen und Implikationen der Fokussierung auf Nebenfolgen bei TA", in: *Technikfolgenabschätzung. Theorie und Praxis*, 16. Jg, 1, pp. 45–54.

Grift, T. (2007): "Robotics in Crop Production", *Encyclopedia of Agricultural, Food, and Biological Engineering*.

Grunwald, A. (2002): *Technikfolgenabschätzung – Eine Einführung*, Edition Sigma, Berlin.

Han, J.; Jo, M.; Park, S. & Kim, S. (2005): "The Educational Use of Home Robots for Children", *IEEE International Workshop on Robot and Human Interactive Communication (ROMAN)*, pp. 378–383.

Heckert, G. & Lenge, R. (2007): "Wenn der Automat am Lenkrad dreht", in: *top agrar. Das Magazin für moderne Landwirtschaft 2*.

Hüttenrauch, H. (2006): *From HCI to HRI: Designing Interaction for a Service Robot*, Stockholm.
Ichbiah, D. (2005): *Roboter. Geschichte-Technik-Entwicklung*, Knesebeck Verlag, München.
IFR Statistical Department (2009): *Serviceroboter in gewerblichen Anwendungen setzen sich durch*, World Robotics, Press release 30.09.2009.
IFR Statistical Department (2008): *World Robotics*.
Landtag BaWü (2007): *Landtag von Baden-Württemberg: Protokoll über die 32. Sitzung vom 11. Oktober*.
Maleri, R. (1997): *Grundlagen der Dienstleistungsproduktion*, Springer, Berlin, Heidelberg. 4. Aufl.
Nelson, R. & Winter, S. (1977): "In search of useful theory in innovation", *Research Policy 6*, pp. 36–76.
Nemet, GF. (2009): "Demand-pull, technology-push, and government-led incentives for non-incremental technical change", *Research Policy 38*, pp. 700–709.
Noack, P. O.; Geimer, M.; Ehrl, M. & Grandl, L. (2010): "Virtuelle Kupplungen von Fahrzeugen – Elektronische Deichsel für landwirtschaftliche Arbeitsmaschinen", *Automatisierung und Roboter in der Landwirtschaft*, KTBL-Schrift 480, pp. 7–16.
Nöcker, G.; Mezger, K. & Kerner, B. (2005): *Vorausschauende Fahrerassistenzsysteme*, Workshop on Driver Assistance Systems, Walting.
NRW (2009): *Mobilität in Nordrhein-Westfalen Daten und Fakten*.
Paschen, H. & Petermann, T. (2005): "Die Institutionalisierung der Technikfolgen-Abschätzung beim Deutschen Bundestag – Ein kurzer Blick zurück", in: Petermann, T. & Grunwald, A. (eds.): *Technikfolgen-Abschätzung für den Deutschen Bundestag. DAS TAB – Erfahrungen und Perspektiven wissenschaftlicher Politikberatung*, Edition Sigma, Berlin, pp.11–18.
Reichert, C. (2009): "Können wir Roboter lieben?", *P.M. 08*, pp. 95–98.
Rip, A. & te Kulve, H. (2008): "Constructive Technology Assessment and Socio-Technical Scenarios", in: Fisher, E. et al. (eds): *The Yearbook of Nanotechnology in Society 1*. Springer Netherlands, pp. 49–70.
Robins, B.; Dautenhahn, K. & Dickerson, P. (2009): "From Isolation to Communication: A Case Study Evaluation of Robot Assisted

Play for Children with Autism with a Minimally Expressive Humanoid Robot", 2nd *Int. Conf. on Advances in Computer-Human Interactions ACHI'09.*

Ruckelshausen, A. (2010): "Autonome Feldroboter", in: *Automatisierung und Roboter in der Landwirtschaft*, KTBL-Schrift 480, pp. 146ff.

Schraft, R. D.; Hägele, M. & Wegener, K. (2004): *Fraunhofer IPA: Service Roboter Visionen*, München, Wien.

Schraft, R. D.; Hägele, M. & Wegener, K. (1993): "Fraunhofer IPA: Service Robots: The Appropriate Level of Automation and the Role of Users/Operators in the Task Execution", *Proceedings of International Conference on Systems Engineering in the Service of Humans, Systems, Man and Cybernetics*, vol. 4, pp. 163–169.

Schreckenberg, M. (2007): "Stau und Panik", *BpB*: *Aus Politik und Zeitgeschichte (APuZ)*, pp. 29–30.

Statistisches Bundesamt (2006): *Im Blickpunkt: Verkehr in Deutschland 2006*, Wiesbaden.

Trevelyan, J. (1999): "Redefining robotics for the new millennium", *The Internat J of Robotics Research 18(12)*, pp. 1211–1223.

United States Senate (1972): "Office of Technology Assessment Act", *Public Law*, pp. 92–484.

Van Lente, H. & Rip, A. (1998): "Expectations in Technological Developments: An Example of Prospective Structures to Be Filled in by Agency", in: Disco, C. & van der Meulen, BE. (eds): *Getting New Technologies Together: Studies in Making Socio-technical Order*, Walter de Gruyter, Berlin, New York.

Weber, J. (2006): "Der Roboter als Menschenfreund", *c't*, Heft 2, pp. 144–149.

Wehmayer, Ch. & Ritter, H. (2010): *Roboter. Was unsere Helfer von morgen heute schon können*, Berlin.

Ethical Aspects of Autonomous Systems

Herman T. Tavani

Abstract: The present essay, organized into three main sections, examines some ethical aspects of autonomous systems (ASs). In the first section, ASs are contrasted with alternative systems that either are made possible or enhanced by developments in artificial intelligence (AI). The second section identifies and briefly examines three ethics-related issues affecting ASs: (moral) agency, autonomy, and moral responsibility. In the third and final section, a framework that builds on the insights of Philip Brey (2004), James Moor (2008), and others is defended as a comprehensive scheme for analyzing ethical issues involving ASs. Whereas the main purpose of this essay is to clarify some conceptual and ethical dimensions of key controversies associated with ASs, as well as to propose a methodological framework for evaluating these controversies, no attempt is made to provide either an in-depth or a definitive account of the issues examined.
Key words: agency, autonomy, autonomous systems, ethical frameworks, moral responsibility
Abbreviations: AAs = artificial agents; AI = artificial intelligence; AmI = ambient intelligence; ASs = autonomous systems; RAEP = Royal Academy of Engineering Report

Introduction

"Autonomous systems are likely to emerge in a number of areas over the coming decades. From unmanned vehicles and robots on the battlefield, to autonomous robotic surgery devices, applications for technologies that can operate without human control, learn as they function and ostensibly make decisions, are growing. These technologies can promise great benefits.... However [they also] raise a number of social, legal, and ethical issues."

(RAEP 2009: 1)

"[W]e are living in a period of technology that promises dramatic change and in which it is not satisfactory to do ethics as usual...Better ethical thinking in terms of being better informed and better ethical action in terms of being more proactive are required."

(James Moor 2008: 26-27)

A cluster of questions that arise from research and development in autonomous systems (ASs) span two distinct areas of philosophical inquiry: metaphysics and applied ethics. From the perspective of metaphysics, some of these questions challenge our received notions of agency, autonomy, freedom, responsibility, and (more broadly) what it means to be human (in an era that some describe as the dawn of "transhumanism"). From the vantage point of applied ethics, on the other hand, questions raised by developments in ASs challenge our conventional understanding of notions such as privacy, trust, moral agency, moral responsibility, and professional responsibility in software/engineering design. In some cases, questions affecting these two (distinct) philosophical areas overlap; this is especially apparent in ASs-related questions concerning agency, autonomy, and (moral) responsibility.

In the present essay,[1] we focus mainly on *ethical* aspects of ASs, including concerns that affect some broader policy issues needed to guide current and future development in that area of technology; where appropriate, however, we also consider some metaphysical questions that undergird the ethical concerns. We begin by defining ASs, which are distinguished from some earlier forms of "automated" and "intelligent" systems also made possible, or significantly enhanced, by research in artificial intelligence (AI). Next, we identify and briefly describe some ethical challenges for ASs identified in the 2009 Royal Academy of Engineering Report, entitled *Autonomous Systems: Social, Legal and Ethical Issue*s (henceforth referred to as RAEP), which includes a discussion of concerns affecting privacy and trust. In that analysis, we also draw some comparisons to the kinds of privacy threats posed by related (emerging) technologies such as pervasive/ubiquitous computing and ambient intelligence (AmI). In the second section, we focus specifically on the impact that ASs have for three ethical categories: moral agency, autonomy, and moral responsibility. We examine these notions in the context of human-machine interaction. In the third and final section, we briefly examine a critique of the standard or "mainstream" method of applied ethics (Brey 2004) and show why that method is not adequate for analyzing many issues affecting ASs. We also briefly examine three different

---

[1] An earlier version of this essay was presented at the Workshop on Perspectives of Information- and Robo-Ethics, Karlsruhe Institute of Technology (Karlsruhe, Germany), 4 December, 2009. I am grateful to Professors Mathias Gutmann and Michael Decker for inviting me to present this paper at the Workshop, and I wish to thank many of the workshop participants for their helpful comments on various aspects of my presentation. Where possible, I have incorporated their suggestions into the present essay. In composing Sections 2.3, 3.1, and 3.3 of this essay, I have drawn substantially from material in the third edition of my book *Ethics and Technology: Controversies, Questions and Strategies for Ethical Computing* (John Wiley and Sons, 2011).

strategies or approaches used to address ethical controversies involving emerging technologies – viz., schemes that Moor (2008) describes as the "ethics-first," "ethics-last," and "dynamic-ethics" approaches. Building on and incorporating many of Brey's and Moor's insights, we propose a comprehensive methodological framework for analyzing ethical issues surrounding ASs.

1. What are Autonomous Systems?

RAEP (p. 2) describes ASs as systems that can "operate without any human intervention" and which have three key properties – viz., they are (1) *adaptive*, (2) capable of *learning* (as they function), and (3) capable of *making decisions*. According to RAEP, the following kinds of devices and entities qualify as instances of ASs: autonomous robotic surgery devices; robots on the battlefield; unmanned vehicles in military/defense applications (e.g., "drones"); transport systems (in commerce); personal care support systems; and "smart" buildings, homes, apartments. Consider the example of a "smart home" with respect to the many kinds of functions that can be performed without any human intervention. Raisinghani, et al. (2004) describe some features of a hypothetical smart home of the not-too-distant future – in the context of ambient intelligence (AmI), i.e., an emerging technology (similar to ASs) that enables people to live and work in environments that respond to them in "intelligent ways"[2] – via a scenario where a young mother arrives home with her daughter.

> "[A] surveillance camera recognizes the young mother, automatically disables the alarm, unlocks the front door as she approaches it and turns on the lights to a level of brightness that the home control system has learned she likes. After dropping off her daughter, the young mother gets ready for grocery shopping. The intelligent refrigerator has studied the family's food consumption over time and knows their preferences as well as what has been consumed since the last time she went shopping. This information has been recorded by an internal tracking system and wireless communication with the intelligent kitchen cabinets. Based on this information, the refrigerator

---

[2] See, for example, Aarts and Marzano 2003; Brey 2005; and Weber et al. 2005 for similar definitions of AmI. Other authors use the expressions "ubiquitous computing" and "pervasive computing" to describe the technology we refer to here as ambient intelligence. For an account of some of the subtle differences in these technologies, see Tavani 2011.

automatically composes a shopping list, retrieves quotations for the items on the list from five different supermarkets in the neighborhood through an Internet link, sends an order to the one with the lowest offer and directs the young mother there. When arriving at the supermarket, the shopping cart has already been filled with the items on her shopping list. Spontaneously, she decides to add three more items to her cart and walks to the check-out. Instead of putting the goods on a belt, the entire cart gets checked out simply by running it past an RFID transponder that detects all items in the cart at once and sends that information to the cash register for processing." (Brey 2005: 157)[3]

A smart home (or a corresponding smart environment), of course, is only one instance of ASs, and we should note that some earlier technologies have also included functions that anticipated at least some of the capabilities found in ASs. So, we can reasonably ask: How are ASs distinguishable from earlier kinds of automated and automatic systems? In differentiating ASs from alternative systems, RAEP (p. 2) distinguishes four levels of gradation: *controlled systems, supervised systems, automatic systems, autonomous systems*. These gradations are determined by factors such as the level of intervention (in terms of control) by human beings for the systems to function effectively. Accordingly, the various systems appear on a continuum where "controlled systems" allow for most human intervention, to autonomous systems in which no human control is required. Table 1 illustrates the four levels of gradation.[4]

---

[3] A slightly longer version of this passage from Raisinghani, et al. 2004 is cited in Brey 2005: 157.
[4] The descriptions and examples included in this table are taken directly from RAEP, p. 2.

Table 1

Type of System	Level of Human Intervention/Control	Example
controlled system	a system that allows humans full or partial control in carrying out its functions	an engine-management system in a car
supervised system	a system that carries out the instructions of a human operator	a programmed lathe
automatic system	a system that carries out a fixed function without a human operator	an elevator
autonomous system	a system that is adaptive and can learn/make decisions (without human intervention)	a "smart environment"

Much more could be said about each of these systems, including the devices and entities comprising them. As noted earlier, however, our primary focus is on ethical aspects and implications of ASs. RAEP (p. 1) identifies a number of key questions in this area. We briefly describe two such concerns: one affecting privacy, and the other involving trust. With respect to the former, RAEP points out that some ASs allow for detailed recording of personal information. So, people who live in automated apartments (or in the kinds of "smart homes" described above) could have vast amounts of information about them recorded and kept by a third party. This practice can, of course, raise some serious privacy concerns. But how are these concerns different from privacy threats generated by earlier computer-related technologies? Langheinrich (2001), writing from the vantage point of AmI-related technologies, describes a number of privacy threats that he believes are significantly different from those posed by earlier technologies and which also easily apply to ASs. For one thing, he points out that no aspect of our life will be secluded from "digitization," because virtually anything we say, do, or even feel, could be "digitized, stored, and retrieved anytime later." Langheinrich also believes that the kinds of privacy threats posed by technologies such as AmI and ASs are qualitatively different from earlier computer-related privacy concerns because of four features underlying these technologies: ubiquity, invisibility, sensing, and memory application. Unfortunately, we cannot examine these distinctions here in the detail they deserve, since doing so would take us beyond the intended scope of the present essay.[5]

We next turn briefly to another ethical concern identified in RAEP, viz., ASs and trust. The Report asks us to consider whether we can trust ASs, especially those systems that are designed in such a way that they cannot be shut down by human operators, to always act in our best interests? It also questions whether we can we trust ASs to override human decisions in certain kinds of cases, given that a human might make bad choices in a stressful situation – e.g., as a result of panic – where the human override would be problematic (RAEP, p. 3). But how much trust should we place in these systems? Lim, Starker, and Larkin (2008) note that questions about trust can involve more than "rational decision making," based on values of "cost effectiveness or of thinking of and for others." The authors also note that in the context of human-machine interactions, we need to define the scope of the boundaries of "ethical trust" from two perspectives: (a) "man to machine" and "machine to man." We recognize that questions about trust raised by these authors are very important and that many trust-related issues affecting ASs will need to be addressed in the near term. In the present essay, however, we are unable to examine, at least not directly, many of the AS-specific issues associated with trust and ethical trust.[6]

We should note that there are a number of ethical concerns affecting ASs that could be examined, in addition to those affecting privacy and trust. However, we limit our analysis to three ethical categories that affect ASs: agency, autonomy, and moral responsibility. We focus on these three notions, in large part, because they involve features or characteristics that apply not only to humans, but also potentially to "machines" such as ASs – e.g., in the event that ASs could, like humans, qualify as (fully) autonomous moral agents that are responsible for their actions.

---

[5] For a fuller discussion of why Langheinrich believes that these features make a qualitative difference with respect to privacy threats involving earlier computer technologies, see Tavani 2011.

[6] For some in-depth discussions of ethical concerns affecting trust in the context of artificial agents (and, by extention, ASs), see Buechner 2011, Grodzinsky, Miller & Wolf 2011, and Taddeo 2011.

2. Agency, Autonomy, and Moral Responsibility in the Context of ASs

In this section, we examine some key questions affecting: (1) agency (including questions about artificial agency, rational agency, moral agency, etc.); (2) autonomy (and related concerns affecting freedom, control, and technological dependency); and (3) moral responsibility (including questions about *who* is morally culpable for failures, accidents, injuries, and deaths caused by ASs). We begin with a look at some implications that research and developments in ASs have for our conventional understanding of notions such as *agent* and *agency*.

2.1 Agency, Agents, and Artificial Agents

Why should we care whether or not ASs can qualify as agents of some sort? Perhaps one important reason can be found in RAEP (p. 1), which asks whether some kinds of ASs should be regarded as either (a) "robotic people" or (b) machines? The Report notes that if ASs are viewed as instances of (a), they would be subject to blame for faults that occur; on the other hand, ASs would not likely be culpable, if they were regarded simply as examples of (b). Although it might seem odd to talk about ASs as "people," robotic or otherwise, we can reasonably ask whether ASs can be regarded as "agents" of some sort: especially as "moral agents."

We begin by asking a basic question: What, exactly, is an agent? In the philosophical literature, an agent is sometimes construed in terms of "an actor," or more particularly, as "acting in someone's behalf" – e.g., as in the case an author's agent or a professional athlete's agent, each, in this case, acting in behalf of the interests of an author or an athlete. Some refer to agents of this type as "fiduciary agents." Himma (2009: p. 119) notes that the idea of agency is conceptually associated with the idea of "being capable of doing something that counts as an act or action." He also notes that while "an action is a doing," not every "doing" qualifies an action.[7] Others, however, also note that some person's (or, for that matter, some entity's) merely carrying out an act, or merely having the capacity to carry out an act, would not

---

[7] Using this distinction, we could invoke Wittgenstein's classic example (in the *Philosophical Investigations* §611-630) involving the difference between someone's "raising his arm" and that person's arm merely going up. While the latter may qualify as a "doing," in Himma's scheme, it would not qualify as an act.

necessarily qualify the actor as an agent.[8] So, if these critics are correct, an entity's having the capacity to act would not be a sufficient condition for that entity's being an agent.

Philosophers describe many different kinds of agents – i.e., they speak of human agents, rational agents, artificial agents, moral agents, and so forth (see, for example, Ganascia, 2007). Of particular interest to us in our analysis of ASs is the notion of an *artificial* agent (or AA)? Dennett (1987) defines an AA as an automaton. In his view, a computer, or even a thermostat that regulates temperatures, would qualify as an AA.[9] But would these kinds of AAs also qualify as genuine agents, or as what some describe as *rational* agents? For example, does the thermostat decide, i.e., make a rational choice, in the process of raising or lowering the temperature in a house, even if it is "acting" on behalf of someone, say the homeowner – i.e., the (human) agent who set or programmed the thermostat? To be a rational agent, the actor must not only be capable of carrying out an act but also be able to decide whether to carry out the act (i.e., towards accomplishing the specific goal in some means-end situation) and to be able to make a decision about which act to carry out, where more than one option is available. While the kind of thermostat described above would not meet the conditions required to be a rational AA, the kinds of ASs identified in RAEP would seem to qualify as rational agents.

Since ASs, and possibly other kinds of AAs, appear to be capable of making decisions, it would seem that not all rational agents are "human agents." A distinguishing feature separating some kinds of rational AAs from human (or natural) agents is that the latter entities are biologically alive. In what sense, however, is this factor relevant? For example, does it also affect the rational agent's ability to be autonomous? Some philosophers argue that (at least some kinds of rational) AAs, like natural or human agents, can be autonomous.

---

[8] For example, Lloyd Carr (unpublished manuscript, p. 18) defines an agent as an entity that is capable of "making a decision." He also argues that an agent "must have a *goal* and must choose from at least two *options* what to do to achieve it" (p. 526, Italics Carr). In this scheme, an agent must be capable of making decisions involved in achieving either its own goal or some goal on the behalf of others (i.e., as in the case of a fiduciary agent).

[9] Originally noted in Ganascia 2007: p. 114. Ganascia also points out that for Dennett, an agent is "an intentional system, i.e., an entity to whom people are inclined to attribute beliefs, desires, will, emotions, etc."

Floridi and Sanders (2004), for instance, define an autonomous AA as an entity that is able to: (i) interact with its environment, (ii) change its internal state dynamically, and (iii) adapt its behavior based on experience. Of course, this view of AAs is controversial and it has been challenged by some who question whether AAs can have the kinds of interests and desires that we tend to associate with someone's capacity for autonomy. We return to the question of whether AAs can meet the threshold for autonomous behavior in our analysis of autonomy in Section 2.2.

We should also note that not every rational agent, including every human agent, can qualify as a *moral agent*. According to Eshleman (2001), the concept of moral agency is generally believed to be limited to a class of beings whose behavior is "subject to moral requirements" and "open to responsibility ascriptions."[10] Floridi and Sanders, who contrast what they call "artificial moral agents" with artificial entities that are simply "moral patients," believe that at least some autonomous AAs can satisfy the conditions for moral agency. But, as in the case of their claim that AAs can be autonomous, many also disagree with Floridi and Sanders' view that AAs can be moral agents. For example, Himma suggests that AAs cannot be moral agents because they lack consciousness, which he argues is required for free will and intentionality. And Johnson (2006) argues that while AAs can be "moral entities," they cannot be moral agents because they do not have intentionality. However, we will not pursue that debate here. We return to the question of whether AAs can be moral agents in Section 2.3.

Even if ASs cannot qualify as moral agents, they could nevertheless be viewed as *social agents*. According to Carr (p. 19), social agents "make decisions that are intended to promote the common interests, the common good of the group." Because individual (component) agents in a complex AS can act together in a cooperative manner to achieve one or more common goals, they would seem to satisfy the requirements for being social agents. Of course, not all social agents necessarily carry out actions that have moral import. And even in cases where a social agent's action has moral implications, it does not necessarily follow that the social agent is responsible in a moral sense for the consequences of its actions. More will be said about social

---

[10] Originally noted in Himma 2008.

agents below in connection with our discussion of ASs as multi-agent systems.

Although we leave open for now the questions of whether ASs can be autonomous agents and (perhaps more importantly) whether they can be moral agents, we argue that ASs do qualify as agents of some sort – viz., they can be *artificial, rational, and social agents*. We further claim that many ASs fit the category of what some AI researchers call "multi-agent systems." That is, ASs can qualify as systems that are *multi-agent* in the sense that (a) they are made up of multiple (individual) agents or sub-agents that constitute a system (of agents), and (b) these agents and sub-agents can interact with one another (as well as with humans) across multiple computer systems and distributed networks. The concept of a multi-agent system used here builds on a standard definition of an AA introduced by Subrahamanian, et al. (2000: 4), in which artificial software agents (i) consist of a collection of software that provide useful services that other agents might employ, and (ii) have the capacity to interact with other agents, both artificial and human, in different environments, ranging from cooperative to hostile.[11]

At least some ASs can also be viewed as "distributed agents" in the sense that the individual agents, and possibly even components of those individual agents (i.e., sub-agents), can be distributed across multiple systems. As such, ASs can have distributed software and hardware components, which, in turn, can be agents. These complex and distributed multi-agent systems can also be social agents (defined above) in the sense that they are members of a ("social") group in which "decisions of the members [can be] collected or pooled together to arrive at a collective decision" (Carr: 19).[12]

Thus, we can now answer the question posed at the beginning of Section 2.1 by noting that ASs are indeed agents. That is, they qualify as agents that are artificial, rational, and multi-component or complex); some can also qualify as distributed agents and as social agents. However, we are not yet able to say whether any kinds of ASs can also qualify as autonomous agents, and, perhaps more importantly, whether they can be moral agents.

---

[11] For an analysis of multi-agent systems that have been used in experiments analyzing relationships of trust between agents, see Buechner and Tavani 2011.

[12] Carr also points out that "each agent's decision contributes to a collective decision process going on in the group."

Ethical Aspects of Autonomous Systems

Even if it turns out that ASs cannot qualify as (full-blown) moral agents, we have seen that the actions of at least some ASs can have a moral impact. Moor (2006) describes the various kinds of ethical impacts that computer systems that "do our bidding as surrogate agents" can have. Although he does not refer to these agents as ASs per se, he believes that the consequences (and potential consequences) of what he calls "ethical agents," can be differentiated into four levels: Ethical Impact Agents, Implicit Ethical Agents, Explicit Ethical Agents, and Full Ethical Agents. He notes that whereas ethical-impact-agents (i.e., the weakest sense) will have ethical consequences to their acts, implicit-ethical-agents have some ethical considerations built into their design and "will employ some automatic ethical actions for fixed situations."[13] While explicit-ethical-agents will have, or at least act as if they had, "more general principles or rules of ethical conduct that are adjusted and interpreted to fit various kinds of situations," full-ethical agents "can make ethical judgments about a wide variety of situations" and in many cases can "provide some justification for them." In Moor's scheme, full-ethical-agents have the kind of ethical features that we usually attribute to ethical agents like us (i.e., what Moor describes as "normal" human adults), including consciousness, free will, and intentionality.

Moor does not claim that either explicit- or full- ethical agents exist or that they will be available anytime in the near term. However, he does suggest that it would be prudent for us to establish clear and explicit policies to address the potential impacts that these various kinds of agents can have. In Section 3, we propose an ethical framework to approach policy issues affecting developments in ASs. Next, however, we try to clear up some earlier questions and concerns that arose in our discussion of "autonomous" behavior in the context of AAs.

2.2 Individual (Human) Autonomy and Autonomous Agents

In this section, we examine two autonomy-related questions, which also overlap at certain points: (1) Are ASs genuinely *autonomous* agents (i.e., do they meet the necessary and sufficient conditions required for autonomous behavior)? (2) What impact can the

---

[13] Moor 2006: 19 notes that a "robotic camel jockey" (i.e., a technology used in Qatar) is an instance of an ethical-impact agent, whereas an airplane's automatic pilot system would be an example of an implicit ethical agent.

development and use of ASs have for our notion of human autonomy (especially at the level of individual autonomy and freedom)? Even if our answer to (1) turns out to be indeterminate or inconclusive, some important questions regarding (2) still need to be considered (and ideally resolved).

What, exactly, is meant by "autonomy"? Although it is an important notion in discussions involving democracy and freedom, *autonomy* is difficult to define. Philosophers have appealed to a wide range of concepts in attempting to provide an adequate account of autonomy. For example, Dworkin (1988: 6) notes that autonomy has been equated with concepts such as liberty, dignity, individuality, responsibility, self knowledge/knowledge of one's interests. He also believes that the only constant feature in these various associations is that autonomy is seen as a "desirable quality."[14] A slightly different account of autonomy can be found in Faden and Beauchamp (1986: 7), whose work in the area of informed consent links autonomy with notions such as "privacy, voluntariness, self-mastery, choosing freely, choosing one's moral position and accepting moral responsibility for one's choices."[15] O'Neill (2002: 29) describes individual autonomy as a "capacity or trait that individuals have to a greater or lesser degree, which they will manifest by acting independently, in ...appropriate ways."

O'Neill points out that while the notion of autonomy can be traced back to antiquity, the conception of *individual autonomy* is more recent. She notes that in ancient Greece, autonomy referred not to individuals but to cities that made their own laws – e.g., an autonomous city was contrasted with a colony that had its laws imposed upon them.[16] Perhaps O'Neill's description of autonomous city states can help us to see how it may be possible to describe collective entities or "collectivities" (i.e., entities other individual humans) as being autonomous in a way that does not sound odd. Thus, it might seem plausible to speak of (at least some types of) ASs as having the capacity for autonomy. However, we can also ask: In what sense of "autonomous" can ASs be autonomous agents? Recall

---

[14] Originally noted in O'Neill 2002: 21–22.
[15] Originally noted in O'Neill 2002: 22.
[16] O'Neill notes that in the modern period, Kant saw autonomy as fundamental for morality, and she points out that contemporary views of autonomy see it as more fundamental to individual *agents*, rather than to morality.

Floridi and Sander's three criteria for an AA to be autonomous (that we identified in the preceding section). Are those criteria reasonable? A critic could point out that while it might make sense for a system (such as a state or political entity) to be viewed as autonomous because it is an aggregate of individual humans (i.e., human agents), who themselves are capable of being autonomous, the same would not hold for a system composed solely (or perhaps even partly) of artificial (i.e., non-human) agents.

A critic could also point out that it might make sense to ascribe autonomy to artificial agents only if those agents were sentient or at least were capable of experiencing desires. Although attempts have been made in the field of affective computing to build in artificial emotions into complex computer systems, it is unclear whether artificial emotions of this sort would be sufficient to qualify for the kinds of desires and interests that some argue are necessary for human autonomy. Unfortunately, in this essay, we must leave open the question of whether ASs are, or can be, fully autonomous – i.e., the first of the above two questions posed at the beginning of this section. Instead, we focus on second question by asking what kind of impact ASs can have on our understanding of human autonomy, especially at the level of individual human agents.

Can ASs enhance human autonomy by giving humans more control over their lives? According to O'Neill (p. 29), "greater autonomy gives individuals greater control" over (a) the ways they live, and (b) their capacities to resist others' demands and institutional pressures. In this view, the notion of autonomy can be closely associated with philosophical concepts of freedom and control. When applied to ASs, however, at least three additional questions arise: (i) How much autonomy should humans be willing to give up to enjoy some of the promised benefits of ASs? (ii) Can humans gain freedom by delegating control to ASs? (iii) On balance, will human autonomy and freedom be enhanced or diminished by the use of ASs?

One might assume, initially at least, that humans will gain more control in AS environments with which they interact because ASs will be more responsive to their needs. But Brey (2005) notes a paradoxical aspect of this claim, when applied to AmI (ambient intelligence), pointing out that "greater control" is presumed to be gained through a "delegation of control to machines." He also suggests, however, that is tantamount to the notion of "gaining control by giving it away." Brey considers some ways in which control can be gained in one sense, and lost in another. With respect to humans *gaining control* as a result of emerging technologies, he notes the human environment may become more controllable because it can: (a) become more responsive to the voluntary actions, intentions, and needs of users; (b) supply humans with detailed and personal information about their environment; and (c) do what people want without having to engage in any cognitive or physical effort. So, as Brey notes, a technology such as AmI (and by extention ASs) has the potential to enhance freedom through its ability to expand certain aspects of our control over the environment – e.g., by freeing us from many routine and tedious tasks that require either cognitive or physical effort.

However, Brey also notes that this kind of technology has the potential to limit freedom significantly and thus can diminish the amount of control that humans have over their environments. For example, he points out that a user could lose control when smart objects perform autonomous actions that do not solely represent his or her interest – i.e., a "smart object" could include a user profile or knowledge base that also is designed to take into account the interests of third parties (especially commercial interests).[17] Brey also suggests that human freedom and autonomy could be undermined if humans become too dependent on machines for their judgments and decisions. Dependency on machines and technology can take many forms. For example, RAEP notes that these can include dependency for performing many of our "cognitive" or "routine" functions as humans, as well as many of our ordinary day-to-day tasks that are viewed as "mundane," "dangerous," or "dirty." We have come to depend a great deal on technology, especially on information/computer technology, in conducting many activities in our daily lives. In the future, humans could develop a dependency on ASs in ways that exceed our current dependency on computing devices. Consider that while future ASs may relieve us of having to worry about performing many of our routine tasks,[18] which we may consider tedious and boring, they could

---

[17] See Brey 2005 for a more thorough discussion of how AmI (and, by extention, ASs) can both enhance and limit human control.

[18] Because humans sometimes vary their routines to avoid monotony, they can be "creative" (in a non-artistic or non-aesthetic sense) by deliberately choosing not to follow a specific routine at a given point. But because an AS would be unaware of a

also relieve us of much of the cognitive kinds of activities that, in the past, have enabled us to feel fulfilled and to flourish as humans.

What would happen to us, as humans, if we were to lose critical cognitive capacities because of an increased dependency on ASs? Consider a precautionary tale from a literary work first published more than 100 years ago. In his classic work *The Machine Stops* (1909), E. M. Forster portrays a futuristic society in which humans have transferred control of much of their lives to a global Machine, which is capable of satisfying their physical and spiritual needs and desires. In surrendering so much control to the Machine, however, people eventually lose touch with the natural world. After a while, the Machine breaks down; unfortunately, however, no one remembers how to repair it.[19] Could the kind of hypothetical scenario described by Forster be realized in the future, where individuals would no longer be required to perform routine cognitive tasks, and instead depend solely on ASs to carry out those functions for them? If so, what would happen if the energy sources that powered the ASs were suddenly lost? Because this worst-case scenario *could* happen, we can reasonably ask whether we have a moral obligation to ourselves not to become too dependent on ASs?

2.3 Moral Responsibility

As noted above, RAEP questions whether (at least some) ASs should be regarded as "robotic people," as opposed to machines, and that this distinction would make a considerable difference with respect to the locus of moral responsibility in cases where the use of ASs results in accidents, injuries, and deaths. We have already argued that ASs can be viewed as agents – viz., as rational agents, social agents, and multi-agent systems in which individual "sub-agents" in the system can be distributed across one or more computer networks. But we have remained silent on the question of whether ASs can qualify as moral agents that can be held responsible for their actions and thus culpable

---

human's desire to engage in an alternative form of behavior (as opposed to carrying out the routine task) at a certain time, the AS could, unwittingly, diminish that human's creativity and freedom. For example, an AS might assume that the human agent would never want to deviate from the set of routine tasks that it has "learned," or inferred, from the human's behavior. This possible scenario for a way in which an AS could limit human creativity and freedom was pointed out to me by Lloyd Carr.

[19] See Forster 1995.

in a moral sense. If ASs themselves cannot be held morally responsible for their actions, then who should? In attempting to make sense of these questions, we begin by briefly examining the concept of moral responsibility.

Philosophers typically discuss the concept of responsibility in connection with conditions such as agency and *causality*. So an agent, A, is held responsible for some act, B, if A caused B. A can also be held *morally* responsible for B even when A (was the cause of B but) did not intend for B to happen. For example, if A causes a brush fire through careless behavior and if the fire spreads in a way that results in the death of someone, C, then we can hold A morally responsible for C's death (even though A did not intend C's death).

With respect to assigning moral culpability for an action, the condition of *intent* can apply even when an agent fails to carry out an intended act. For example, consider a scenario in which X, a disgruntled employee, intends to blow up a corporation's computer lab, but at the last minute is discovered and prevented from doing so. Even though X failed to carry out his objective—i.e., causing the bomb to detonate in the computer lab—we hold X morally culpable because of his intentions.

Does it make sense to hold ASs responsible in a moral sense – i.e., to say that they are culpable – for their actions? Consider the case of an autonomous transport system (a type of AS) that causes, or significantly contributes to, a highway accident involving multiple automobiles with human passengers. Should the AS be held morally responsible for any deaths or injuries that result? If not, then who, if anyone (or any entity), is responsible in a moral sense? RAEP (p. 4) notes that laws require that someone, i.e., some "person," be held responsible for an accident that results in injury or death. In a legal sense, however, a person can be non-human. For example, an entity such as corporation can qualify as a "legal person." But can a "computer system," such as an AS, also be viewed as a *person* for legal purposes? And even if it could, would it also follow that we could attribute blame or praise (in a moral sense) to the AS?

With respect to the question of whether we can ascribe moral responsibility to artificial agents (and, by extention, to ASs), two different kinds of strategies have been proposed to show that we could. One strategy argues that we need to separate the concept of agency from responsibility (Stahl 2006), while the other suggests that we need to expand the notion of "moral agent" (Floridi 2007). We begin with a brief description of the first strategy. Stahl (p. 112) notes that moral responsibility is usually discussed in terms of personhood and personal characteristics, which he believes computers may or may not have. But because personal characteristics are closely tied to agency, he suggests that we should separate the question of responsibility from agency. Stahl then argues that "responsibility ascriptions" can be made to computers if we use an alternative notion of responsibility as a kind of "social construct," which he calls "quasi responsibility." In Stahl's scheme, quasi responsibility redefines traditional responsibility by emphasizing the "social aims of responsibility ascriptions." His argument, which is fairly complex and thus cannot be examined here in the detail that it deserves, suggests that certain kinds of computer systems may meet the requirements of his expanded criteria for ascribing moral responsibility.

We next briefly describe the second of the two strategies – i.e., Floridi's claim that we need to expand the notion of "moral agent." Arguing that the concept of moral agent needs to be broadened to include legal persons as well as natural ones, Floridi begins by noting that entities such as partnerships, corporations, and governments are all instances of "legal persons." But Floridi points out that our concept of moral agency is still "human based," even though actions that have moral consequences are no longer limited to individual human agents. He also points out that the legal entities other than individual (human) persons are, in effect, based on "aggregates of humans or persons." Although aggregates of human agents can qualify as legal persons, the same does not apply to non-human agents such as artificial agents. However, Floridi believes that because some artificial agents, like non-human legal persons, are also "sources of moral actions" (i.e., have an ethical impact), they could be held morally responsible in some sense. And because some of these agents also can have roles that are distributed over computer networks, Floridi argues that we need a concept of "distributed morality."

Both Stahl's and Floridi's views are a controversial, and they cannot be elaborated upon (or developed) here, since an adequate analysis of their arguments would take us beyond the scope of this essay. Nevertheless, both arguments raise interesting points with respect to why it could make sense to attribute moral responsibility to some kinds of computer systems because of the kinds of actions they can perform.

Let us assume for purposes of this essay that ASs cannot meet the threshold for being moral agents and thus cannot be morally responsible for any accidents, injuries, and deaths that result from their use. Even if we hold that assumption, two important questions still need to be addressed: (1) If ASs are not morally responsible agents, which agents (i.e., individual humans and groups) should be held accountable for injuries and deaths that result from their use – i.e., the software programmer(s)/engineer(s), system designer(s), manufacturer(s), user(s), or some combination of them? (2) If these groups are not morally responsible, in the strict sense of that term, should they still be held legally accountable – i.e., liable in a legal sense – for actions that result from accidents, injuries, and deaths involving ASs?[20]

Nissenbaum (2007) suggests that we distinguish between responsibility and accountability, noting that "responsibility" is only part of what is covered by the "robust and intuitive notion of accountability." In Nissenbaum's scheme, accountability is not only a broader concept than responsibility, it also means that someone, or some group of individuals, or perhaps even an entire organization is *answerable*. Nissenbaum further notes that in a computing context, such as ASs, accountability "means that there will be someone, or several people, to *answer* ... for malfunctions that cause or risk grave injuries".[21]

In spite of the fact that we are increasingly dependent on safety-critical and life-critical systems controlled by computers, Nissenbaum believes that the notion of accountability has been "systematically undermined" in the computer/information era. She argues that a major barrier to attributing accountability to the developers of safety critical-software systems (such as ASs) is because of the fact that "many

---

[20] We can distinguish responsibility from the related notions of liability and accountability. Responsibility differs from liability in that liability is a legal concept, sometimes used in the narrow sense of "strict liability." To be strictly liable for harm is to be liable to compensate for it even though one did not necessarily bring it about through faulty action. Here, the moral notion of blame may be left out. A property owner may be legally liable for an injury to a guest who falls in the property owner's house, but it does not necessarily follow that the property owner was also morally responsible for any resulting injury.

[21] Nissenbaum 2007: 274. [Italics Added]

hands" can be involved in their development. Because complex computer systems such as ASs are typically developed in large organizational settings, and because they are the products of engineering teams or of corporations, as opposed to the products of a single programmer working in isolation, many individuals (or many "hands") are involved in their development. Thus it is very likely that no single individual could grasp all of the software code used in developing one of these systems.[22] As a result, it is difficult to determine who exactly is accountable whenever one of these systems results in personal injury or harm to individuals.

When thinking about the problem of many hands from the perspective of strict moral responsibility, as opposed to accountability, Nissenbaum notes that two types of difficulties arise: First, we tend to attribute moral responsibility for an accident to an individual, but not to groups or "collectivities."[23] The second difficulty arises because the concept of moral responsibility is often thought of as something that is *exclusionary*: In other words, if we can show that A is responsible for C, then we might infer that B cannot be held responsible for C, and vice versa. However, Nissenbaum suggests that in using the concept of "accountability" instead of "responsibility," we can circumvent this difficulty. For example, she notes that holding A accountable for making unauthorized copies of proprietary software does not necessarily preclude our also holding B accountable as well, as when B pays A to make unauthorized copies of the software. So Nissenbaum believes that holding one individual accountable for some harm need not necessarily let others off the hook, because several individuals (and organizations or collectivities) may be accountable. According to Nissenbaum, "We should hold each fully accountable because many hands ought not necessarily lighten the burden of accountability."

We have seen that in Nissenabum's scheme, the notion of *accountability* is a broader concept than responsibility; it is also non-exclusionary, and it can apply to groups as well as individuals.

---

[22] Ibid. p. 275. Here, Nissenbaum also includes additional guidelines such as "independent auditing" and "excellent documentation".
[23] Thus we sometimes encounter difficulties when we try to attribute blame to an organization. Nissenbaum suggests that by using "accountability" we can avoid the tendency to think only at the level of individuals in matters typically associated with assigning moral responsibility.

However, we cannot infer that she intends for it to apply to computer systems such as ASs. So, we conclude this section by leaving open the question of whether ASs can be full-blown moral agents and thus responsible, even in part, for the consequences of their actions. But we have seen that multiple agents (including humans, and collectivities such as corporations and governments) can all be accountable, either solely or partly, and either directly or indirectly (in contributory ways), for actions involving ASs that result in damage, personal injuries, and deaths.[24]

3. An Ethical Framework for Analyzing Controversies in ASs

In the preceding sections, we focused on some conceptual and metaphysical questions underlying key ethical issues affecting agency, autonomy, and moral responsibility. Unfortunately, we had to leave open the answers to many of these questions. However, we also saw that because of some practical implications involving many of these questions, it would be prudent for us to frame some clear and explicit policies to address certain challenges posed by ASs. On the one hand, it may be difficult to fill some of the "policy vacuums" generated by ASs without first resolving the corresponding "conceptual muddles" (Moor 2007) surrounding them, some of which were identified and examined in our discussion of key questions in Section 2. Nevertheless, we need to resolve as many of the AS-related policy issues as we can, since plans for implementing this technology are currently underway. Arriving at a methodological framework that would enable us to accomplish this task is the main objective of Section 3.

3.1 The "Standard Methodology" for Applied Ethics

We begin this section by looking at the standard methodology used in research in applied ethics to see whether it is adequate for analyzing concerns that arise in ASs. Brey (2004) notes that the standard method

---

[24] Although ASs themselves might not be morally responsible agents, it is possible that they could be held morally accountable in Nissenbaum's scheme in the sense of moral rectification – i.e., they could be identified as one of the "accountable parties" in situations of redress involving monetary damages. However, we cannot pursue this point here.

has three distinct stages, which require the researcher to: 1) identify a particular controversial practice as a moral problem; 2) describe and analyze the problem by clarifying concepts and examining the factual data associated with that problem; and 3) apply moral theories and principles in the deliberative process in order to reach a position about the particular moral issue.[25] However, Brey also argues that this model is not adequate for analyzing issues in computer ethics; so, by extention, it would also seem to follow that he would also find the model inadequate for analyzing issues affecting ASs.

Before examining Brey's critique, we should note that he is not the first to question the so-called standard methodology. In fact, many philosophers have pointed out difficulties with one or more of the three stages involved. For our purposes, however, the most important critique can be found in Moor (2007), who suggests that the second stage in the method – i.e., clarifying concepts – may be more difficult than it initially appears. Moor points out that in the case of computer technology, which is "logically malleable," many "conceptual muddles" arise that need to be resolved[26] before we can move from the second to the third stage.

We agree that Moor provides an important insight, which also requires us to revise or expand upon the conditions described at the second stage of the standard applied-ethics method. But are the revisions suggested by Moor sufficient to resolve Brey's concerns? Brey believes that a critical problem for the method also often occurs at the first stage. He acknowledges that although the standard model might work well in many fields of applied ethics, such as medical ethics and business ethics, it does not always fare well in computer ethics (even with the kinds of revisions suggested by Moor). Specifically, Brey argues that the standard method, when used to identify ethical aspects of computer technology, tends to focus almost exclusively on the *uses* of that technology. As such, the standard method fails to pay sufficient

---

[25] See Brey 2004: 55–56.

[26] Moor also notes that sometimes these conceptual confusions need to be resolved before we can fill the vacuum of policies – i.e., by either framing new policies or revising existing ones. We should note that Moor does not claim that conceptual muddles and policy vacuums arise only in computer ethics, as opposed to other areas of applied ethics. But Moor (2001) does claim that we can find proportionately more policy vacuums in this area of applied ethics because of the "logical malleability" of computer technology, which makes possible new forms of behavior for which we have either no (clear and explicit) policies or inadequate policies.

attention to certain features that may be embedded in the technology itself, such as design features, which may also have moral import.

Brey argues that computer technology has certain built-in values and biases that are not always obvious or easy to detect, and he worries that these biases can easily go unnoticed by researchers.[27] And because many researchers also tend to assume computer technology is inherently neutral, Brey proposes a method that causes us to question this assumption. He argues that we first need to locate what he calls "embedded normative values" in the technology.

He also points out that the standard method of applied ethics tends to focus on "known moral controversies" and that, as a result, it fails to identify certain practices involving a technology that also can have moral import but are not yet known. Brey refers to such practices as having "morally opaque" (or morally nontransparent) aspects, which he contrasts with "morally transparent" features that tend to be easily recognized as morally problematic. For example, he notes that many people are aware that the practice of placing closed circuit video surveillance cameras in undisclosed locations is controversial from a moral point of view. But Brey also notes that it can be much more difficult to discern morally opaque features in technology. These features can be morally opaque because either (a) they are known but perceived to be morally neutral, or (b) they are unknown.[28]

Consider an example of a morally opaque feature in which a technology is well known. Most Internet users are familiar with search-engine technology. What users might fail to recognize, however, is that certain uses of search engines can be morally controversial with respect to personal privacy. Consequently, one of the features of search-engine technology can be morally controversial in a sense that it is not obvious, or transparent, to many people, including those who are very familiar with and who use search-engine technology. So while a well-known technology, like search engines, might appear to be morally neutral, a closer analysis of practices involving this technology will reveal that it also has moral implications.

---

[27] Brey's work in this area has been influenced by the Value Sensitive Design (VSD) approach used by Batya Friedman and others. For a description of the VSD model, see Friedman, Kahn & Borning 2008.

[28] For more detailed account of this notion of morally opaque features, see Brey 2004: 56–57.

Next, consider an example of a morally opaque (or morally nontransparent) feature. Computerized practices involving RFID (Radio Frequency Identification) technology would be unknown to those who have never heard of the concept of RFID (and who are thus unfamiliar with that technology). However, we should not assume that RFID technology is morally neutral merely because it may be unknown to nontechnical people, including many ethicists. Even if information-gathering techniques involving RFIDs are opaque to many users, practices involving the use of RFID technology can raise moral concerns pertaining to personal privacy.

Brey argues that an adequate methodology must first identify, or "disclose," features that, without proper probing and analysis, would go unnoticed as having moral implications. Thus, an extremely important step in Brey's method is to identify moral values embedded in the various features and practices associated with a particular technology by *disclosing* them. We return to Brey's "disclosive" method in the concluding section of this essay.

3.2 Two Approaches in Applied Ethics: The "Ethics First" and "Ethics Last" Models for Resolving Technology-Related Controversies

Many of us will eventually be affected by AS technology, regardless of whether or not we, as individuals, elect to use this technology; in some cases, the deployment of ASs could have significantly negative social consequences for virtually everyone. In fact, Wallach and Allen (2009: 4) predict that within the next few years, a "catastrophic accident" will occur as a result of decision made by a computer system "independent of human oversight."[29] So, it would seem that we would benefit from having a clear and explicit ethical framework in place to guide us in the development and use of ASs. In this section, we briefly examine two kinds of ethical frameworks that have been used vis-à-vis recent technologies: the "ethics-last" and "ethics-first" models.

The ethics-last model is perhaps the standard or traditional model that has been most used by scientists and policy makers in addressing

---

[29] The authors note that in October 2007, a "semiautonomous robotic cannon deployed by the South African army malfunctioned, killing 9 soldiers and wounding 14 others" (Wallach & Allen: p. 4).

ethical aspects of new technologies. Moor (2008) and others view this model as the default, or "business as usual," approach; it is also generally considered to be a "reactive," rather than proactive, model in the sense that ethical considerations have followed scientific developments rather than informing scientific research. Moor notes that in most scientific research areas, ethics has had to play "catch up," in which case the ethical guidelines were also typically developed in response to cases where serious harm had already resulted. The assumption in these cases has been that we could simply wait until a particular technology has been perfected, or at least completed, before we have to begin to worry about its ethical implications. Some defenders of this view seem to suggest that that until we know that a given technology *can do* the kinds of things its proponents and developers claim that it can do, we don't have to worry about any of the ethical impacts of that technology. But critics argue that by this point in the technological development cycle, it may be too late to address the ethical issues in a timely manner.

Many critics who reject the ethics-last approach also tend to assume that the best alternative must be an ethics-first approach. The latter approach to ethics is generally viewed to be proactive; a now classic example of an ethics-first model is the Ethical, Legal, and Social Implications (ELSI) framework used in the Human Genome Project.[30] Before work was able to proceed on that project, the anticipated ethical, legal, and social implications had to be identified and addressed. Specific concerns affecting ethical aspects of genomic research were identified in the ELSI model – e.g., concerns about privacy, confidentiality, fairness, etc. – and specific guidelines for addressing each of these concerns had to be "built into" the scientific research model.

Some have argued that we should look to the ELSI model for clues about how to construct appropriate ethical frameworks to guide research in various emerging technologies. For example, Kurzweill (2005) has suggested that an ELSI-like model should be developed and used to guide research in nanotechnology. By extention, his arguments could also be used to support the adoption of this framework for guiding research in ASs.

---

[30] For more detail, see the ELSI Research Program, established by the National Human Genome Research Institute.

Although many see the ELSI model as a vast improvement over traditional (ethics-last) frameworks, where ethics was mainly an "afterthought," critics have pointed out some flaws in the ELSI model when it is applied to research in emerging technological areas such as bioinformatics and computational genomics, especially in population studies and DNA databases. For example, I have argued elsewhere that data-mining technology and related computational techniques pose a serious challenge for ELSI's effectiveness with respect to meeting privacy and informed-consent concerns in the context of genomic research.[31]

Other critics have pointed out different kinds of problems with the ELSI framework. For example, Moor and Weckert (2004) argue that the ELSI model, like all ethics-first models, has problems because it depends in large part on a "factual determination" of the specific harms and benefits in implementing a given technology before an ethical assessment can be done. But, in the case of some technologies, such as nanotechnology, the authors note that it is very difficult to know what the future will be in five or ten years, let alone twenty or more years. This critique applies in the case of ASs, as well. If we developed an ethics model along the lines of the ELSI framework, it might seem appropriate to put a moratorium on AS research until we get all of the facts. But Moor and Weckert point out that while a moratorium on future research would halt technology developments, it will not advance ethics in the area of that technology. They also argue that turning back to the "ethics-last model" is not desirable either, because once a technology is in place, much unnecessary harm may already have occurred. So, in Moor and Weckert's scheme, neither an ethics-first nor an ethics-last model is satisfactory for analyzing ethical issues that arise in emerging technologies.

3.3 The "Dynamic Ethics" Model

In articulating and defending what they call the "dynamic ethics" model, Moor and Weckert argue that ethics is something that needs to be done continually – i.e., as (a) technology develops, and (b) the technology's potential consequences become better understood.[32] So, in their scheme, ethics has a "factual," as well as a normative

---

[31] See my critique of the ELSI framework in Tavani 2004.
[32] This method is more fully developed in Moor 2008.

component, and the factual/descriptive component needs to be analyzed regularly. It is in this sense that their scheme is "dynamic" (as opposed to static). And it is in the context of having to continually assess a technology and it anticipated consequences in light of the factual data available, as required by the dynamic-ethics model, that Weckert and Moor describe some of the shortcomings of the Precautionary Principle. For example, they note that applying this principle in a technological field can easily result in having to place a moratorium on research in that field.

Other critics have pointed out that the expression *precautionary principle* is ambiguous, because there may be many different formulations of this principle. Clarke (2005) points out that both "strong" and "weak" versions of the Precautionary Principle have been put forth by its proponents. He believes that the strong version of the Precautionary Principle is "too strong," while the weak version is "too weak." Weckert and Moor interpret the Precautionary Principle in the following way:

> "If some action has a possibility of causing harm, then that action should not be undertaken or some measure should be put in its place to minimize or eliminate the potential harms." (Weckert and Moor 2004: 12)

When applying this principle to questions about research in a specific technological area, Weckert and Moor believe that three different "categories of "harm" need to be analyzed: direct harm, harm by misuse, and harm by mistake or accident. Consider that the kinds of risks involved in each type of harm can differ drastically in the use of various technologies, including ASs. So, Weckert and Moor argue that it is important to differentiate among these three kinds of harms in assessing a technology and in framing policies for research and development in a particular technological field, such as ASs.

Critics other than Weckert and Moor have also expressed concerns about the negative impacts that moratoriums can have for the advancement of certain kind of technologies, especially those that may be needed and that may be appropriately safe. For example, RAEP (p. 8) notes that in order "to avoid *stalling a technology* that could be of significant benefit, Government should *engage in* early consideration of regulatory policy so that such systems can be introduced" [italics added]. Moor and Weckert's dynamics-ethics model not only endorses the view expressed in RAEP, it also provides a specific strategy for engaging in early discussions and debates about policy considerations. As we debate whether and how to go forward with research and development in AS technology, we can see how neither an ethics-first nor an ethics-last model is adequate. We agree with Moor (2008) and Moor and Weckert (2004) that it is necessary to establish a set of ethical guidelines that can be continually updated as new factual information about ASs becomes available. Hence, we also applaud RAEP for providing us with some current factual data in this area of technology. RAEP provides us with an excellent start; however, the Report will need to be updated continuously, as factual data surrounding ASs changes and becomes clearer.

We believe that the dynamic-ethics model, as articulated by Moor and Weckert, is the most promising scheme (of the models we examined) for addressing the kinds of policy issues that will likely arise in the near-term development and use of ASs. We also believe, however, that some of Brey's insights regarding the standard method of applied ethics, described in Section 3.1, can be incorporated into a comprehensive ethical framework that can inform the policy debate for ASs.

4. Concluding Remarks: A Comprehensive Ethical Framework for Analyzing ASs

Earlier in this section, we examined Brey's critique of the standard, threefold methodology used in applied ethics. There, we also saw that some moral issues affecting computer-based technologies could be opaque, or non-transparent, and that these issues needed to be identified or "disclosed" before we could proceed. We believe that elements of Brey's "discolsive method" can be integrated into a comprehensive framework, which also incorporated the insights of Moor and Weckert's dynamic-ethics model.

In considering how Brey's disclosive method can be implemented in the context of ASs, we also need to ask which individuals and groups are responsible for carrying out the various steps or functions required. Brey points out that his method is both multidisciplinary and multilevel. It is multi- or inter-disciplinary because it requires that computer scientists, philosophers, and social scientists collaborate. And his method is multilevel because it requires three levels of analysis: (i) disclosure level, (ii) theoretical level, and (iii) application level. First of all, the moral values embedded in the design of ASs must be disclosed. To do this, we need computer scientists and engineers because they understand ASs and related computer technology much better than philosophers and social scientists do. However, social scientists are also needed to evaluate system design and make it more user-friendly. Then philosophers can determine whether existing ethical theories are adequate to test the newly disclosed moral issues or whether more theory is needed. Finally, computer scientists, philosophers, and social scientists must cooperate by applying ethical theory in deliberations about the moral issues affecting ASs in order to propose coherent normative policies.[33]

If we combine the insights in Brey's disclosive method with the insights in Moor and Weckert's dynamic-ethics model, we can revise and improve upon the standard three-step framework of applied ethics described at the beginning of Section 3. In particular, we propose the following comprehensive (four-step) framework for approaching ethical controversies, and for formulating coherent policies, affecting ASs:

1. *Identify* a practice involving ASs, or a technological feature of ASs, that is controversial from a moral perspective.
    1a. Disclose any hidden or opaque features that have moral import (Brey);
2. *Analyze* the ethical issue by clarifying concepts and situating them in the context of ASs.
    2a. Identify any "policy vacuums" and clear up any "conceptual muddles" (Moor) that arise in the context of ASs.
3. *Deliberate* on the ethical issue(s) affecting ASs.
    3a. Apply one or more standard ethical theories – i.e., utilitarianism, deontology,[34] etc. – to the moral issue(s) in ASs, and determine whether any new or revised ethical theories are needed.

---

[33] See Brey 2004: 64–65. Some believe that in the deliberations involved in applying ethical theory to a particular moral problem, an additional process is required. For example, van den Hoven (1997) has noted that methodological schemes must also address the "problem of justification of moral judgments." However, we do not examine that process in the present essay.

[34] Moor (2004) has proposed a comprehensive ethical theory that combines aspect of utilitarianism and deontology, which he calls "Just Consequentialism".

4. *Continue to update* the ethical analysis as ASs develop and their potential social consequences become better understood.

   4a. Differentiate between the factual/descriptive and normative components of ASs.

   4b. Revise the policies as necessary, especially as the factual data or components change or as that information becomes clearer (Moor and Weckert).

   4c. When information about plans for the design and development of newly proposed technological features affecting ASs becomes available, go to Step 1.

In this essay, we have arguably left more questions unanswered than answered. Nevertheless, we have identified some key ethical issues that will likely arise as a result of developments in ASs. We have also described the kinds of impacts these issues could have for our conventional understanding of concepts affecting three ethical categories: agency, autonomy, and responsibility. Finally, we proposed a comprehensive ethical framework that can guide us in addressing policy issues that will likely arise and need to be resolved, as ASs are developed and implemented.

Acknowledgments

This essay has benefited from conversations with Jeff Buechner, Rutgers University, and Lloyd Carr, Rivier College. I am especially thankful to Lloyd Carr for his helpful suggestions on an earlier draft of this essay.

References

Aarts, E. & Stefano, M. (eds.) (2003): *The New Everyday: Views on Ambient Intelligence*, 101 Publishers, Rotterdam, The Netherlands.

Brey, P. (2004): "Disclosive Computer Ethics", in: Spinello, R. A. & Tavani, H. T. (eds.): *Readings in CyberEthics*, 2nd ed. Sudbury, MA: Jones and Bartlett, pp. 55–66. Reprinted from *Computers and Society*, Vol. 30, No. 4, pp. 10–16.

Brey, P. (2005): "Freedom and Privacy in Ambient Intelligence", *Ethics and Information Technology*, Vol. 7, No. 4, pp. 157–166.

Buechner, J. (2011): "Trust, Privacy, and Frame Problems in Social and Business E-Networks, Part 1", *Information*, Vol. 2, No. 1, pp. 195–216.

Buechner, J. & Tavani, H. T. (2011): "Trust and Multi-Agent Systems: Applying the 'Diffuse, Default Model' of Trust to Experiments Involving Artificial Agents", *Ethics and Information Technology*, Vol. 13, No. 1, pp. 39–51.

Carr, L. J. (forthcoming): *Making Good Choices: An Introduction to Practical Reasoning*. Unpublished Manuscript.

Clarke, S. (2005): "Future Technologies, Dystopic Futures and the Precautionary Principle", *Ethics and Information Technology*, Vol. 7, No. 4, pp. 121–126.

Dennett, D. (1987): *The Intentional Stance*, MIT Press, Cambridge, MA.

Dworkin, G. (1988): *The Theory and Practice of Autonomy*, Cambridge University Press, Cambridge, MA.

ELSI Research Program. National Human Genome Research Institute. Available at: http://www.genome.gov/10001618.

Eshleman, A. (2001): "Moral Responsibility", in: Zalta, E. (ed.): *Stanford Encyclopedia of Philosophy*, available at http://plato.stanford.edu/entries/computing-responsibility.

Faden, R.; Beauchamp, T. & King, N. M. P. (1986): *A History and Theory of Informed Consent*, Oxford University Press, New York.

Floridi, L. (2007): "Distributed Morality in Multi-Agent Systems", in: Hinman, L. et al., (eds.): *Proceedings of the Seventh International Conference on Computer Ethics – Philosophical Enquiry (CEPE 2007)*, Enchede, The Netherlands: CCIT Workshop Proceedings, pp. 110–112.

Floridi, L. & Sanders, J. W. (2004): "On the Morality of Artificial Agents", *Minds and Machines*, Vol. 14, No. 3. pp. 349–379.

Forster, E. M. (1995): "The Machine Stops", in: Johnson, D. G. & Nissenbaum, H. (eds.): *Computing, Ethics, and Social Values*, Prentice Hall, Upper Saddle River, NJ, pp. 694–713. Reprinted from *The Eternal Moment and Other Short Stories*, Harcourt Brace, New York, 1970.

Friedman, B.; Kahn, P. & Borning, A. (2008): "Value Sensitive Design and Information Systems", in: Himma, K. E. & Tavani, H. T. (eds.): *The Handbook of Information and Computer Ethics*, John Wiley and Sons, Hoboken, NJ, pp. 69–101.

Ganascia, J.-G. (2007): "Using Monotic Logics to Model Machine Ethics", in: Hinman, L. et al., (eds.): *Proceedings of the Seventh International Conference on Computer Ethics – Philosophical Enquiry (CEPE 2007)*, CCIT Workshop Proceedings, Enchede, The Netherlands, pp. 113–121.

Grodzinsky, F. S.; Miller, K. W. & Wolf, M. J. (2011): "Developing Artificial Agents Worthy of Trust: Would You Buy a Used Car from this Artificial Agent?", *Ethics and Information Technology*, Vol. 13, No. 1, pp. 17–27.

Himma, K. E. (2009): "Artificial Agency, Consciousness, and the Criteria for Moral Agency: What Properties Must an Artificial Agent Have to Be a Moral Agent?", *Ethics and Information Technology*, Vol. 11, No. 1, pp. 19–29.

Johnson, D. G. (2006): "Computer Systems: Moral Entities but not Moral Agents", *Ethics and Information Technology*, Vol. 8, No. 4, pp. 195–204.

Kurzweil, R. (2005): "Nanoscience, Nanotechnology, and Ethics: Promise and Peril", in: Mitcham, C. (ed.): *Encyclopedia of Science, Technology, and Ethics*, Vol. 1., Macmillan, New York, pp. xli–xlvi.

Langheinrich, M. (2001): "Privacy by Design – Principles of Privacy-Aware Ubiquitous Systems", in: *Proceedings of the Third International Conference on Ubiquitous Computing*, Springer-Verlag, New York, pp. 273–291.

Lim, H. C.; Stocker, R. & Larkin, H. (2008): "Review of Trust and Machine Ethics Research: Towards a Bio-Inspired Computational Model of Ethical Trust (CMET)", in: *Proceedings of the 3rd International Conference on Bio-Inspired Models of Network, Information, and Computing Systems*, Hyogo, Japan, Nov. 25–27, Article No. 8.

Moor, J. H. (2001): "The Future of Computer Ethics: You Ain't Seen Nothin' Yet!", *Ethics and Information Technology*, Vol. 3, No. 2, pp. 89–91.

Moor, J. H. (2004): "Just Consequentialism and Computing", in: Spinello, R. A. & Tavani, H. T. (eds.): *Readings in CyberEthics*, 2nd ed., Jones and Bartlett, Sudbury, MA, pp. 407–417.

Moor, J. H. (2006): "The Nature, Importance, and Difficulty of Machine Ethics", *IEEE Intelligent Systems,* Vol. 21, No. 4, pp. 18–21.

Moor, J. H. (2007): "What Is Computer Ethics?", in: Weckert, J. (ed.): *Computer Ethics*, Ashgate, Aldershot, UK, pp. 31–40. Reprinted from *Metaphilosophy*, Vol. 16, No. 4, 1985, pp. 266–275.

Moor, J. H. (2008): "Why We Need Better Ethics for Emerging Technologies", in: Hoven, J. van den & Weckert, J. (eds.): *Information Technology and Moral Philosophy*, Cambridge University Press, New York, pp. 26–39.

Moor, J. H. & Weckert, J. (2004): "Nanoethics: Assessing the Nanoscale from an Ethical Point of View", in: Baird, D.; Nordmann, A. & Schummer, J. (eds.): *Discovering the Nanoscale*, IOS Press, Amsterdam, The Netherlands, pp. 301–310.

Nissenbaum, H. (2007): "Computing and Accountability", in: Weckert, J. (ed.): *Computer Ethics*, Ashgate, Aldershot, UK, pp. 273–280. Reprinted from *Communications of the ACM*, Vol. 37, 1994, pp. 37–40.

O'Neill, O. (2002): *Autonomy and Trust in Bioethics*, Cambridge University Press, Cambridge, MA.

Raisinghani, M.; Benoit, A.; Ding, J.; Gomez, M.; Gupta, K.; Gusila, V.; Power, D. & Schmedding, O. (2004): "Ambient Intelligence: Changing Forms of Human-Computer Interaction and Their Social Implications", *Journal of Digital Information*, Vol. 5, No. 4, Article No. 271, pp. 08–24.

Stahl, B. C. (2006): "Responsible Computers? A Case for Ascribing Quasi-Responsibility to Computers Independent of Personhood and Agency", *Ethics and Information Technology*, Vol. 8, No. 4, pp. 205–213.

Subrahamanian, V. S.; Bonatti, J.; Dix, J.; Editor, T.; Kraus, S.; Ozcan, F. & Ross, R. (2000): *Heterogeneous Agent Systems: Theory and Implementation*, MIT Press, Cambridge, MA.

Taddeo, M. (2011): "Modelling Trust in Artificial Agents, A First Step in the Analysis of E-trust", *Minds and Machines*, Vol. 20, No. 2, pp. 243–257.

Tavani, H. T. (2004): "Genomic Research and Data-Mining Technology: Implications for Personal Privacy and Informed Consent", *Ethics and Information Technology*, Vol. 6, No. 1, pp. 15–28.

Tavani, H. T. (2011): *Ethics and Technology: Controversies, Questions, and Strategies for Ethical Computing*, 3rd ed., John Wiley and Sons, Hoboken, NJ.

The Royal Academy of Engineering Report (2009): *Autonomous Systems: Social, Legal and Ethical Issues*, London. Available at: www.raeng.org.uk/autonomoussystems.

Van den Hoven, J. (1997): "Computer Ethics and Moral Methodology", *Metaphilosophy*, Vol. 28, No. 3, pp. 234–248.

Wallach, W. & Allen, C. (2009): *Moral Machines: Teaching Robots Right from Wrong*, Oxford University Press, New York.

Weber, W.; Rabaey, J. & Aarts, E. (eds.) (2005): *Ambient Intelligence*, Springer, New York.

Weckert, J. (2004): "Lilliputian Computer Ethics", in: Spinello, R. A. & Tavani, H. T. (eds.): *Readings in CyberEthics*, 2nd ed., Jones and Bartlett, Sudbury, MA, pp. 690–697. Reprinted from *Metaphilosophy*, Vol. 33, No. 3, 2002, pp. 366–375.

Weckert, J. (2006): "The Control of Scientific Research: The Case of Nanotechnology", in: Tavani, H. T. (ed.): *Ethics, Computing, and Genomics*, Jones and Bartlett, Sudbury, MA, pp. 323–339. Reprinted from *Australian Journal of Professional and Applied Ethics*, Vol. 3, 2001, pp. 29–44.

Weckert, J. & Moor, J. H. (2004): "Using the Precautionary Principle in Nanotechnology Policy Making", *Asia Pacific Nanotechnology Forum News Journal*, Vol. 3, No. 4, pp. 12–14.

Wittgenstein, L. (1953): *Philosophical Investigations*, Trans. G. E. M. Anscombe, Macmillan, New York.

Anthropocentric-Based Robotic and Autonomous Systems: Assessment for New Organisational Options[1]

António Brandão Moniz

Abstract: Research activities at European level on the concept of new working environments offers considerable attention to the challenges of the increased competencies of people working together with automated technologies. Since the decade of 1980 the development of approaches for the humanization of work organization, and for the development of participative organizational options induced to new proposals related to the development of complex and integrated automated systems. More recently, the debate also covers issues related to working perception of people dealing with autonomous systems (e.g. Autonomous robotics) in tasks related to production planning, to programming and to process control. In fact, today one can understand the wider use of the anthropocentrism concept of production architectures, when understanding the new quality of these systems. In this paper is analysed the evolution of these concepts related to governance of ICT applied to manufacturing and industrial services in research programmes strengthening very much the 'classical' concept of anthropocentric-based systems. It is emerging a new value of the intuitive capacities and human knowledge in the optimization and flexibilization of the manufacturing processes. While this would be a pre-condition to understand the human-robot communication needs, there is also a need to take into consideration the qualitative variables in the definition and design of robotic systems, jobs and production systems.
Keywords: working environment; work organization; Autonomous robotics; governance; human-robot communication

1. Introduction

There are considerable research activities at European level on the concept of new working environments. These activities encompass the challenges of the increased competencies of people working together with automated technologies, and especially with robots. In fact, 344 thousand industrial robots were installed in European factories by the end of 2008. It represents a huge 'population' of machines that had its

---

[1] Text based on the paper presented at the Conference "Autonomous systems: inter-relations of technical and societal issues" held at Monte de Caparica (Portugal), Universidade Nova de Lisboa, November, 5th and 6th 2009 and organized by IET-Research Centre on Enterprise and Work Innovation under the collaboration project CRUP/DAAD on "Technology Assessment of Autonomous Robotics" of FCT-UNL and ITAS-KIT. The author wish to thank the constructive comments of Bettina-Johanna Krings, although the responsibility of the article, remains by the author.

strong "demographic" increase in the 1990 decade, and still is been used intensively in manufacturing industry. One may surely state, that this sector is the one where most automated systems and robotics have been applied. Many application fields, within the debate on robots, still are not realised in industry, or not use in larger extend in the service sector.

This 'population' of industrial robots is used mostly for handling operations, or for welding and dispensing, but also for processing and for assembling and disassembling[2]. According to the European Platform of Robotics, one can consider different type of robots: industrial robots (for working environments as 'workers', 'co-workers' or for logistics), professional service robots (all applications), domestic service robots (co-workers, logistics and surveillance robots), security robots (the same as for domestic service, plus exploration and inspection), and even space robots (the same for industrial robots, plus exploration and inspection)[3].

The increase of industrial robots took part in the last decades at the same time when the work organization in the manufacturing industry has been under strong restructuring process. In several sectors (automotive, metal, electronics etc.) such restructuring has used the introduction of microelectronics in the labor process to improve the productivity and flexibility. In some case it had also used the industrial robots as one of the main technologies to support the renovation or upgrading of value chain. This modernization implied a new mode in terms of qualification needs and – above all – new organizational alternatives. In this chapter we will discuss the evolution of the technology options for new market conditions within the strategic choices governed by the management models in the manufacturing industry. These choices can be done among the more technocentric approaches (supported on Tayloristic "one best way" model) or the anthropocentric ones based on idea that a more participative and learning organization is the one that can cope with flexibility and complexity of technical systems.

---

[2] An important task that had become a recent increase because of the development of the recycling industry.

[3] We will not consider for propose of this chapter the "edutainment" (education and/or entertainment) robots, although some autonomous systems have been experienced with such application of service robots (e.g. "robocup").

2. Research Findings on Working Conditions and Automation

Already in the 1970 decade, the International Labour Office (ILO) was the main institution that published several studies about the relation between workers and technology, specially, ICT and embedded micro-processors in the working environment (most were mentioned by other sociological research publications, like Braverman 1974; Bell 1976; Kern & Schumann 1984; Piore & Sabel 1984; Bessant 1989), and came to the result that the increase of automated system have created new types of problems. Without any doubt, such new technologies increased the pace of work and the intensitivity of human tasks, but in other cases it gave floor to the implementation of new forms of work organization and increased participation in the decision making by workers. In both cases, most of the problems definitely seemed related with new needs for the improvement of working conditions.

On the other side automated technologies offereda new opportunity to include the need for expertise from the workers to better balance the production lines and also improved the quality control of what? At that time (decade of 1970 and early 1980) some studies were developed on automated transfer systems in the manufacturing industry, the development of numerical controlled machine tools, and the beginning of the introduction of micro-processors in the office work (Brandt, 1992; Brödner, 1990; Davis and Wacker, 1982, Hertog and Schroder, 1989). They gave a new insight on the importance of social aspects related with the introduction of ICT at the workplace. Tese studies raised the attention to new aspects related with the work organization, and with knowledge needs to deal with such technologies (specific training, basic competences, informal knowledge).

Besides ILO, in this decade, the European Foundation for the Improvement of Living and Working Conditions [4] was founded as a European Commission unit to analyse issues related to the emergence of new forms of work organization, and for the analysis of working condition. It has supported and published several sectoral studies held during these first years of the decade of 1980. The European Foundation also started in 1992 the organization and publication of

---

[4] It was one of the first units to be established to work in specialised areas of EU policy. Specifically, it was set up by the European Council in 1975, to contribute to the planning and design of better living and working conditions in Europe.

European surveys on work environment (European Foundation, 1992). Later, during the years 1993-98, it carried out a major programme of research dealing with the nature and extent of the direct participation that is at the heart of new forms of work organization, known as EPOC (Employee direct Participation in Organisational Change) programme. Some years later (in 2001), the European Foundation set up the European Monitoring Centre on Change (EMCC) which is an information resource established to promote an understanding of how to anticipate and manage change. This observatory had the full support of the European Parliament, the European Commission and the social partners. These activities were focused on the organizational changes and anticipation of work environment changes. Some studies on automation and robotics were held but no case studies were developed. Mostly they were integrated in more general reports on national or sectoral restructuring and modernisation processes. The aim of those studies was to understand the sector restructuring in Europe. Because of that some comments, observations and data collection was made relating to examples of automation development in some member states, and sectors. Some of those studies also included analysis about robotics implementation in manufacturing (automobile, electronics, metal engineering, etc.). Just few public discussions were held specifically on those topics of technology change and automation implications. And those held were especially focused on the role of social partners in the restructuring process. No specific results can be retrieved on the issue of robotics and/or autonomous systems used in the productive sectors.

On another institutional basis, the European Commission coordinated also research activities during the 1980s in the field of Anthropocentric Robotic Systems that influenced the ESPRIT-European Strategic Programme of Research in Information Technologies programme during several decades, and a wide group of European social scientist. It was a field that encompassed the wider topic of human-centred systems, or advanced production systems and participative organisation, but more focused on robotics. The attention to such field does not arised only after 2000 with the so-called "Lisbon Strategy" but from decades earlier, for example, with the activities at the Forecasting and Assessment in Science and Technologies (FAST) unit of DG Research. This unit paved the ground for new networks and research projects (Brandt 1992; CEC-

FAST 1987; FAST 1989; Hertog & Schroder 1989; Jones, Kovács & Moniz 1988; Kidd 1992). Such projects and research networks contributed to the knowledge towards the different technological design options (agile manufacturing, balanced automation systems, virtual enterprises, production networks, etc.). And, although many studies were published on social aspects of automation, these underlined the dimensions of technological design alternatives.

Usually, it was assumed that robotic technology was "given" and developed by advanced engineering centres using the most advanced concepts and state-of-the-art knowledge of technology. Only on the base of these preconditions it was possible to understand the results of implementation, the impacts or the implications for employment, new qualification needs for the workers or the whole job design. Thus, after those (awareness) studies it was possible to understand how important the design of technology became, and how far it can be driven by political, ethical or social aspects. Alt least, it also was understood that different industrial robots manufacturers develop their products according to different organizational principles or approaches. For example, in the D. Brandt report on anthropocentric production systems experiences (Brandt 1990) are mentioned experiences of machine tool manufacturers that developed their systems to facilitate the operators' control of production process. More recently, in the SME Robot project (see pages…) a generic graphic human-machine interface was developed where the processes are combined and inserts process parameters according to user description. That means that no manual configuration effort is required or no knowledge of device interface is required. It can use the graphical layout to generate a real robot controller system with assistance to help the user to correctly set-up the robot cell. Comau, KUKA, ABB and Reis industrial robot manufacturers were the companies involved in such projects. This means they can also profit from the results. But also this means that the research outcomes can be an important tool for an organizational alternative where robot operators can participate in the production control process.

In this sense the FAST unit on 'forecasting and assessment on science and technology' assumed as research field the "anthropocentric production systems", and paved the ground for new networks and research project. In particular, that happened within some of the first

ESPRIT projects.[5] These projects were based on the assumption of the feasibility of design and implementation of flexible manufacturing systems (FMS) and computer integrated manufacturing (CIM) systems that use human operators as key elements of such automation strategies (cf. Brandt 1992; Brödner 1990; Kidd 1992). Key elements because they were designed not to exclude the human participation on operative tasks, and also on planning and programming [6]. On the contrary, these strategies intended to use and to integrate human operators' skills and competences in order to improve the decision processes of workers and robot operators in shop floor manufacturing environment: Either in the product design phase or during the manufacture process. In this case, that automation strategy of increased human involvement in the decision process was done where automated machinery need to be programmed. And especially when there are higher risk probabilities in terms of quality assessment and control in complex manufacturing environments.

As mentioned by Rauner and Ruth "the method of user participation is based on the assumption that the involvement of the users will cause better systems, because on the one hand it better meets the needs and skills of the working people, on the other hand only the users at shop floor level have the knowledge of the 'real' production processes which of course must be included in the technical design process. Evidently users must be involved from the beginning and during the whole participative process" (Rauner & Ruth 1991: 21). This is still valid in the present days.

Since the European 5th Framework Programme of R&D, that is to say only from 1999 onwards, new projects have been supported to develop some specific concepts and ideas, like "participatory technology assessment", "work process knowledge", learning organisations, collaborative knowledge modelling, or "virtual

---

[5] Especially the ESPRIT 1217/1199 project on "Human-centred CIM Systems" that was pioneering the organised research at the EC level on these issues). The project ESPRIT 534 (Development of a Flexible Automated Assembly Cell and Associated Human Factor Study), also was focused on the same topics. More information on this can be also read at Nichols & Jones 1994; Laessoe & Rassmussen 1989; Burns 2000.

[6] Some companies were announcing about the future "unmanned factory", about the total automated units that did not need or use human work operation. That would be the highest achievement in terms of competitivity. Such naïve approach produced strong impacts at the management structures.

organisations", among others. That means such research projects could use new concepts of management sciences and integrate the major experiences and results from the work organization restructuring models (semi-autonomous working groups, production islands, "U" assembly lines, autonomous cells, multitasking working places, etc.). The discussion over international experiences of new forms of work organization in the manufacturing sector (especially in the automobile, chemical and electronic companies) were progressively integrated into new research programs on ICT engineering, or in the social sciences agenda.

These new concepts were rooted into the organizational approach of socio-technical design based on the Emery, Trist or Gustavsen studies in the decade of 1960. The Tavistock Institute research findings from the 1950 decade were being used by the new organizational research approaches almost up to four decades later. Under that European research framework programme, some projects were dealing with flexible work practices based on principles such as decentralization, multi-competences, vertical and horizontal integration of tasks, participation and co-operation, that were already features of human-centred approaches to automation systems. This was also the case for TSER [7] program projects like SOWING [8].

This was also the case for the engineering research on automation and development of manufacturing integration through ICT components and design of new productions flows that could improve the working conditions of the workers and on the same time to improve the productivity levels.

When taken these aspects into account it is necessary to analyse, and to assess these integrated socio-technical system approaches. Several were the experiences to develop such integration, specially the new systemic relation between the organization, the technical system and the social and economical environment. Some cases have been mentioned by Clegg & Corbett 1987, Brandt 1992, or Rauner & Ruth 1991, but also by authors from the socio-technical approach, like Child or Mumford (cf. Castillo 1988), or others more disperse from the sociological and management sciences literature.

---

[7] TSER stands for Targeted Socio-Economical Research and was the sub-programme of R&D financial support for social sciences research projects in the 4th and 5th European Framework Programme.

[8] http://www.uta.fi/laitokset/tyoelama/sowing/sowing.html.

These experiences were in the same stream of the research on alternative organization of work. They followed the main findings from social research in Europe, butalso in the US or in Japan [9], where they pointed to the emergence of network-based information economy with an intense restructuring process on the level of manufacturing organizations. That led to new technological needs, and also to new social and economic demands. Quality, productivity, flexibility, uncertainty, complexity, efficiency were concepts that seem contradictory, but they could be tackled integrating simultaneously within a social and a technological dimension. Some of those experiences at the manufacturing level were related to the design of new robotics cells and integration of those cells into highly sophisticated manufacturing systems that still could use the participation of human decision skills in production planning, programming and control. At least there was some effort to integrate human participation into technological advance.

Nowadays one can recognize that the demands for the improvement of the working circumstances have been cooled down by competition limits and by an intense growing of work pressure and employment instability. With such processes of degradation of work conditions, also the push for a technological development of social design of autonomous technical systems seems in a stabilized process. It has been only focused on intelligent ambiance and machine-machine communication systems[10]. Some research is also supporting the inclusion of human tacit knowledge into artificial reasoning with more powerful programming tools. But this human-machine interfacing is basically instrumental, and not a social or political dimension of the technical design option.

3. Experiences with Anthropocentric Strategies for Automation and Robot Systems in Manufacturing

Most of the experiences on anthropocentric strategies for automation in manufacturing had their floor in Europe. As Rauner and Ruth

---

[9] The CAPIRN project (Culture and Production International Research Network) developed the concept of "industrial culture" from case studies from the major industrialized countries, and their outcomes were also an input to the FAST program on anthropocentric production systems (cf. Rauner & Ruth 1991).
[10] Cf. Ribeiro and Barata 2006 or Moniz 2007.

underlines, that "the implicit 'Eurocentric' orientation (...) finds expression, for example, in the welfare state premises included that do not exist to the same degree in the US and Japen" (Rauner & Ruth, 1991: 7). They followed the industrial approaches in Japan to participatory design of organizations and implementation of quality control policies in sectors where was needed a major involvement of human operators. The most studied ones were held in Sweden (Volvo experiences) and Denmark (B&O, MAN B&W Diesel and other companies in metal and electronics sectors), all involving strongly the social partners in the restructuring process. But other studies were held in Germany and UK (all in wider ranges of sectors [11]), and some others in other industrialized countries like Italy (most in Emilia Romagna, either related with automotive or electronics, and in other Northern Italy regions with garment and textile sector), Spain (in the Mondragón region) or France (in several different sectors and regions)[12]. Here the main issue for its application was to cope with problem solving in productivity and flexibility of production systems. When compared to the same type of organization in the US and in Japan, the European companies were much left behind in terms of productivity capacities. The fact that they could not achieve the flexibility capacities of Japanese and US firms in the same sectors that mean that the results in terms of productivity were also poor.

In Japan the participative strategies were developed and applied since late 70s, and in the US the lean production methods were applied in late 1980s in the manufacturing sectors. In Europe, only the Scandinavian experiences are based on these technical systems (ILO, 1984). As it was described above, robotic systems in manufacturing has been considered (in the decade of 1980 by ILO, by OECD or the Vienna Centre) as a technology responsible for wealth and higher standards of living in Europe, not only due to higher levels of attained productivity, but also to the contribution to improve working conditions. That was, at least, the main argument for the increased acceptance of the introduction of robotic systems at the workplace.

---

[11] Brandt (1992) mentions cases of Thyssen and Hoesch (steel, Germany), Keller, Felten&Guillaume and Lubos&Bayer (metal engineering, Germany), Rolls-Royce and Westland Helicopters (aircraft, UK), Lucas Engineering (electronics, UK).

[12] It is worth to consult most of the references presented in this chapter. In particular, Clegg & Corbett 1987; Castillo 1988; Warner, Wobbe & Brödner 1990; Rauner & Ruth 1991; Brandt 1992; Kidd 1992; Lehner 1992; Freyssenet 1995; Durand, Stewart & Castillo 1998, or even Valeyre 2009.

And manufacturing industry was without doubt the sector where most of robots and automated systems were applied and developed.

Basically, experiences like the Volvo Kalmar case, the SAAB, or the MAN B&D Diesel, or many others from Norway and Finland were related to the implementation of new forms of work organization. But in almost all of the manufacturing industry cases, the implementation of robots and flexible manufacturing systems was done smoothly with the participation of work council and the workers directly involved in the restructuring processes

The Scandinavian socio-technical systems involve self-managed teamwork and work enrichment by multi-skilling. Learning organisations are characterised by strong individual and collective learning dynamics in the workplace, notably with regard to problem-solving activities related to unforeseen events such as dysfunctions in production and with regard to innovation processes. These organisations need high levels of autonomy, initiative and communication at work on the part of employees and attach great importance to autonomous teams and project teams. Based on collective reflexive returns to tasks and events and assigning a larger intelligibility to work (Freyssenet 1995), they clearly break with Taylorist principles (cf. Valeyre et al. 2009).

Jürgens, Malsch and Dohse (1993) argued that the high average levels of qualified labour in the automotive and sectors or in companies that introduced robotic systems, was an argument to develop the experiences with work organization and to design the robotic systems in order to integrate the cooperation of human tasks.

The technical and economical advantages that follow such experiences are associated to improved quality (less rejects and flaws) and increased responsiveness. However, it can also induce shorter throughput times, lower indirect costs and an easier planning and control of production processes. The development of organizational innovations with flexible automation systems imply simplified material flows than with conventional organizational models, and also implies smaller production areas and swifter response to quantitative and qualitative changes in demand. Less breakdowns and increased capacity for innovation and continuous (productivity, quality) improvement, are also features of those systems.

These experiences have shown that even from the social and human point of view the benefits of implementation robotic systems can be considered as an increasing quality of working life, a higher job satisfaction through meaningful rewarding tasks, and a higher degree of motivation and involvement. It implies also a greater personal flexibility and adaptation, and an improved ability, creativity and skills of the shop floor personnel, which requires higher levels of qualification.

An enriched direct interpersonal communication and social relations, increased collective and co-operative will, and a greater capacity for collective learning of new practices are also human and social benefits of those above mentioned systems that articulate organizational innovation and flexible technology applied to production. Thus, in the very beginning of the scientific debates based on research programmes about robotic systems had a positive impact on the emergence of knowledge about the organizational aspects related to the implementation of advanced automated and integrated systems. Such knowledge made possible further research on organizational conditions to provide a better usage of robotic systems and advanced integrated automation. Several experiences were supported and reported through European projects to study these new forms of work organization with automated systems.

In the next item we will focus more on the European research frameworks and the projects on robotics that included dimensions related to user-interfaces, to new hardware configurations to face user needs, or to new software strategies centred on human (normally, shop floor operator competences, not engineer or technician competences).

4. The Research Frameworks in Europe and Robotics

In fact, since the decade of 1990, automation and robotics were at the stake of large European research projects. Several of these projects developed anthropocentric automation approaches. Most of them included inter-disciplinary research teams (engineers, sociologists, management scientists, computer scientists and social psychologists) and provided very interesting scientific literature on major issues related with the challenges that manufacturing industry was facing Europe by the end of the 20th century. With the emergence of new innovation problems (globalization, network and virtual enterprises, technical integration) the focus was becoming more technical-oriented. Although, one can find in every European framework programme of RTD a *continuum* of projects that are dealing with the human-centred configuration of automated systems. New approaches were tested, new more complex experiences took place with the support of those programs, and the debate could continue. From the beginning of the new century, the Lisbon strategy offered also new topics to be responded, and the research institutions together with industry firms tried to cope with those new issues of the framework programmes. In the next lines we report on that evolution and on project examples under each of those European programmes.

Since the 2nd Framework Programme, some ESPRIT and BRITE-Basic Research in Industrial Technologies for Europe projects can be considered as reference frameworks for the collaborative research between the Computer Sciences, Quality and Production Engineering, and Sociology. It was under these projects that social sciences could have major applied research in manufacturing environments [13]. Later, in the ESPRIT 4 programme [14] the research activity on robotics was focusing in four domains: a) "Integration in Manufacturing" (IiM), b) "High Performance Computing and Networking" (HPCN), c) "Technologies for Components and Subsystems" (TCS), and d) "Long Term Research" (LTR). But these domains of 4th Framework Programme included also the issue "user-centred development" on robotics. Such User-Centred Development issue included the integration of user-centred approaches into methods and tools supporting the design and development of systems. Also could be defined through the concept of "Usability Support Environments" which means that it should support user's involvement and feed back through techniques and tools such as early story-board prototyping, simulations to evaluate user reaction, user profile analysis, and so forth.

---

[13] In particular projects with special references to Social Sciences can be mentioned, as the project ESPRIT 1199/1217 "Human-Centered CIM Systems", or the project ESPRIT 5564 "Integrated Design and Evaluation of Assembly Lines within CIM", or the BRITE projects 1381 (on interactive knowledge based shop floor control systems), 3302 (on Decision Support Systems) or 3345 (on flexible production groups), or even the ESPRIT exploratory action 5603 on "Joint Technical and Organizational Design of CIM systems for SME's". Some of these issues were already discussed in a previous article (Moniz 2007) on the importance of these projects for the emergence of such techno-organisational concepts.

[14] This European strategic programme was held from 1994 until 1998 by the European Commission involving all the member states.

This new European programme was focusing much more the ICT research towards the usability principles and the human-machine interfaces improvement. However, the organizational issues related to job design in complex and integrated systems was not anymore a research topic. The IiM domain should be the one where such topics should be developed under R&D projects.

An overview of Robotics Technologies in RTD Programmes of the European Community under the 4th Framework Programme was published by the European Commission by Skordas and Robrock, and there they specify that the domain IiM also focused on robotics projects and preparatory support and transfer activities that are specific to the manufacturing domain. These are related to the theme of 'Intelligent Production Systems' and Equipment comprising some research tasks (what is that?). Among those it seems worth to mention the development of enhanced man-machines IT interfaces for control systems and shop-floor control (mostly in manufacturing industry), and the development of distributed computing environments supporting novel control and decision support methods, for control of manufacturing processes. IT became a clear dominant technical role within the technical systems and specifically in the last decades the European RTD Programmes in the field of autonomous systems (including robotics) were supporting almost exclusively interfaces systems. But also research tasks were taken under that Integration in Manufacturing domain, as the development in IT components and subsystems and embedded micro-devices, and their integration, for open, intelligent, autonomous mechatronic systems. This implies another field that also requires possible user-centred strategies is the integration of real-time quality and performance monitoring functions in flexible manufacturing systems. But no research projects were supported in these fields.

One of the IiM project clusters was the "Intelligent Equipment and control" that comprises a total number of 10 ESPRIT projects in the areas of enhanced man machine interfaces for shop-floor control, computing environments for control of manufacturing processes and IT components and subsystems and embedded micro-devices for robotics and mechatronic systems. But again, no projects appeared to develop further knowledge on the relation of organization and technology, beyond these areas of human-machine interfaces or control systems.

The sectors of robotics manufacturing and machine communications could be represented through the projects AMIRA (EP22646) and RACKS (EP20468) that were the single projects focusing on aspects related to users of this automation technology. The objective of AMIRA was the development of the next generation of advanced man-machine interfaces. Also it was intended to support tools to end-users of robot manipulators for efficient application of robot systems and robotised workcells. The RACKS project was concerned with the situation in the field bus based market and tackles the bottleneck of heavy dependence of manufacturing systems towards the underlying technology used of communication networks. It had also the aim to develop standard user-level common interfaces rendering application programs compatible with a wide range of system architectures. In another ESPRIT field, the Industrial and Materials Technologies Programme (IMT) of the 4th Framework Programme replaced the former BRITE-EURAM, but continued to include a research agenda in the robotics field, covering several topics: intelligent assembly, mechatronics, and micro-system technologies, new quality oriented intelligent and flexible production systems, tele-operated multifunctional robotic systems, joining, inspection and repairing systems incorporating mechatronics, micro-systems, sensors and actuators for real time adaptive control and research on new automatic control and systems theory concepts. Also in this production technologies are of IMT it included the field of "human and organisational factors in production systems", but no projects reflecting such field was supported.

In April 1997 the European Commission published its Green Paper on "Partnership for a New Organisation of Work" (European Commission 1997). As Brödner and Latniak mention, "it did not really produce a signal for departures to new frontiers; it was rather turned down instead during public debates that followed. In the time after, a Communication Paper entitled 'Modernising the Organisation of Work – a Positive Approach to Change' (European Commission 1998) was issued in November 1998, and in March 1999 the European Work Organisation Network (EWON) has been established. These initiatives signed the weight the Commission assigned to the theme. Yet, their impact on the further development of new forms of work organisation has been rather low so far, although the Network appears to be necessary and helpful for improving the knowledge base across the member states, for exchanging experiences, and for raising public awareness for work organisation issues" (Brödner & Latniak 2002: 7). In fact, we can find two reasons for this contradictory situation:

a) The European Green Paper is published by the end of the 1990 decade when the expectations on the organizational innovation reach the highest level. Many publications and experiences have shown that the participative and learning organizations could increase the productivity and product quality where the technological requisites have shown complexity and high modular integration;
b) The last decade (from early 2000) could be characterized by an intensification of labour in a process of increased segmentation of the value chain at a global level. This socio-economical trend pressured technological innovation into a decrease of costs and standardization of production processes.

Such contradictory trends have shown that it could be possible to increase production levels with a decrease of labour costs, with higher levels of control and flexibilisation. That implied a continuous investment on automation with more complex human-machine interfacing for a more reliable manufacturing control and management process. In the European 5th Framework Programme the involvement of larger companies in larger projects was envisaged, and new technological needs were under test. The programme IST (Information Society Technologies[15]) was again the most financially supported programme among all European RTD activities and had a specific topic on robotics: Beyond Robotics. The conclusions from Robotics Working Group Meeting 2002 of the IST programme reflected the main problems to be found in the technology field of robotics until the last years, like the following[16]:

- Interaction with robotic systems is extremely important as system only will be considered as "good" as their interface with the user.
- Today simple brain interfaces have started to emerge.

---

[15] It was not anymore called as ESPRIT programme.
[16] Cf. ftp://ftp.cordis.europa.eu/pub/ist/docs/fet/fetro-28.pdf.

- Today (simple) multi-modal interfaces do exist for interaction with robots. Interfaces are either highly constrained, non-robust and/or require significant training.
- This call for significant advances in both, sensory perception, multi-modal interaction, methods for extended dialogue behaviours and integration of "physical behaviour" with the more traditional interaction modalities.
- A significant problem in design of robotic systems has been the lack of flexible and robust perception system that allows the system to operate in unconstrained environments.
- There is thus a need for careful consideration of the fusion of sensory perception beyond traditional semantic/Bayesian methods.

Again the topics related with the human user were based on "interface with the user", or "brain interfaces", training needs, "extended dialogue behaviours". But a concept about job design for operators of such systems was still missing. All efforts were based on the software and hardware aspects of robotics, but none on the integration robotic systems in "real life" environments in manufacture industry. Such topics revealed also the necessary developments of this technology. In fact, in recent years, the "usability" of robotic systems and the interface with their operators became a central issue for the research and the development of most used robotic technology, but the stress was put on software dimensions. The so called "brain interfaces" are recognized to be still in an early phase of the concept development. But other interfaces have been the main research topic in the recent years. Especially when related to distributed computing and large integration of sensors. Examples from the Karlsruhe experiences on autonomous robotics show us the evidence of such trends.
Later, at 6th Framework Programme (FP6) was approved the Robotics Platform (http://www.robotics-platform.eu/cms/index.php) as one of the European Technology Platforms-ETP supported. This European Robotics Technology Platform (EUROP) was founded in 2005, but in fact its roots go back to October 2004, when leading European robotics organisations started to formulate the need for a consolidated approach to European robotics. As the other ETP, the platform EUROP is an industry-driven framework for the main stakeholders in robotics to strengthen Europe's competitiveness in robotic R&D, as

well as global markets, and announced that it should contribute to improve quality of life.

In this European RTD programme the project that we can mention is the SMErobot (http://www.smerobot.org/). In this project there was an intention to empower the supply chain of robot automation by focusing on the needs and culture of SME manufacturing with regard to planning, operation and maintenance. That could be developed through a robot development capable of understanding human-like instructions (by voice, gesture, graphics), to increase the safety and productive human awareness in a shared space with robots (using cooperative principles, and not using protection fences). It is clear that in future robot instruction schemes it will be required the use of intuitive, multimodal interfaces and preferably human communication channels, such as speech and gestures. A strong effort has been made in this field for the last twenty years. Identification and localization of work pieces, automatic generation or adaptation of programs and process parameters are also required for minimizing programming efforts. In this project was concluded that the absence of highly skilled robotic programmers meant that relatively easy tasks take an average of 40 hours of programming for the average SME. The aim would be that robot programming should be as simple as telling a colleague to perform a certain task. That was also an aim of the first anthropocentric robot systems. The SMErobot project provided guidelines for anyone developing interfaces for industrial robots as how to design multi-modal interfaces based on voice, gesture or manual guidance for natural and intuitive human-robot interaction. That was the main objective to overcome the mentioned limits in the development of such systems to be applied to manufacturing companies, and especially to SME. It is a coordination of several European activities that understand the usage of these autonomous systems much more than only IT programming systems that try to establish simplistic forms of reasoning to be easily understood by humans. The aims behind this research and development programme are grounded on artificial intelligence concepts and tools that can be articulated with social needs and competences requirements at the SME level.

Another project (PHRIENDS) was financed under this same framework programme and is about developing key components of the next generation of robots, including industrial robots and assist devices (http://www.phriends.eu/project.htm). This includes robots for the non-industrial applications market (service, health-care, and entertainment), and they were designed to share the environment and to physically interact with people. Such machines have – under this European project – to meet strict safety standards. The project faced new challenges to the design of all components of the robot, including mechanics, control, planning algorithms and supervision systems. It was envisaged an integrated approach to the co-design of robots for safe physical interaction with humans. That means to design robots that are intrinsically safe, and control them to deliver performance. Also financed under the FP6 was PACO-PLUS project (http://www.paco-plus.org/). This project brought together an interdisciplinary research team to design and build cognitive robots capable of developing perceptual, behavioural and cognitive categories that can be used, communicated and shared with other humans and artificial agents. This European project is undertaking the development of an integrated robotic system with humanoid traits to support interaction with people, in other words, to build a complex anthropomorphic robot. The researchers mention that "anthropomorphism is desirable because it makes interaction easier and also supports the transfer of ideas from psychology and neuroscience to robotics" (project webpage), although this idea is not proved or based in any evidence.

The ETHICBOTS is a project with some links to social sciences that aimed at identifying techno-ethical case-studies on the basis of a state-of-the-art survey in human-machine integration based on Robotics, Bionics, and AI for IA. This project is trying to identify and analyze techno-ethical issues concerning these forms of human-machine integration by reference to case-studies analysis. Will also establish a techno-ethically aware community of researchers, by promoting workshops, dissemination, training activities, and the construction of an internet knowledge-base and generate inputs to EU for techno-ethical monitoring, warning, and opinion generation. It was not clear if social scientists were participating directly in the project integrated in research teams with engineers and computer scientists.

Finally, at FP7 the euRobotics - Coordination Action for Robotics in Europe consortium covers the complete robotics picture including industry, service (both professional and domestic), security and space with the following aims: a) to develop a Strategic Research Agenda in

Europe; b) address the broader impact of Advanced Robotics on society assessing the legal, social and ethical issues surrounding the introduction of Advanced Robots that directly interact with their users in everyday human environments. It will also assess the educational issues. It includes key players from both the industrially driven EUROP network and the academic network EURON. This Framework Programme is still running and new projects can tackle aspects that were previously pointed as needed to go deeper in the research. The technology assessment of these new systems can reach now a new standard level and integrate new tools for following up the project outcomes and to define new fields for further research. It is still too soon to evaluate those outcomes.

And under the FP7 the EU Project LIREC (LIving with Robots and IntEractive Companions)[17] seeks to establish a multi-faceted theory of artificial long-term companions. Is also an aim to embody this theory in robust and innovative robotics and in technologies of autonomous systems. It is also intended to experimentally verify both the theory and technology in real social environments, and to address social, psychological and cognitive foundations and consequences of such technological artifacts entering our daily lives.

HUMOUR is an EU-funded research project (http://www.humourproject.eu/) at the FP7 which aims at investigating and developing efficient robot strategies to facilitate the acquisition of motor skills. It tries to develop robot agents based on an advanced understanding of human euro-motor control, its development and skill acquisitioin. It aims also to extend the domain of Brain Computer Interface (BCI) technologies to the fields of motor learning and neuro-motor rehabilitation. Based also in the human-robot communication process, the CommRob project (http://www.commrob.eu/) has an underlying assumption concerning the robot's interaction design that it should be based on principles of human-human communication in order to provide an interface that is intuitive and easy to use. The development of the communication platform envisioned in this project also provides another research challenge related to the dialogue design, namely that dialogue models should be generic and reusable. The research was oriented to design dialogue models based on established principles for human-machine interaction and ensuring

---

[17] More information can be found at http://www.lirec.org/.

that these models are thoroughly evaluated in realistic usage situations.

As one can understand, these projects were focusing along the last years several concepts associated to anthropocentric approaches, although sometimes in a very limited way. The examples of those concepts are the following:

a) design intrinsically safe robots, and control them to deliver performance (Phriends project)
b) integrated robotic system with humanoid traits to support interaction with people (Paco-Plus project)
c) techno-ethical issues concerning these forms of human-machine integration (Ethicbots project)
d) innovative robotics and autonomous systems technologies for human interaction (Lirec project)
e) development
of robot agents based on an advanced understanding of human neuro-motor control (Humour project)
f) intuitive robot's interaction design based on principles of human-human communication (CommRob project)

These were some few projects among a large database of European projects on robotics and autonomous systems [18]. It is possible to retrieve the projects that are dealing with social, human, ethical or legal aspects. The result is only these above mentioned projects where one can have a stronger impression on the research concepts that are been supported within the most important (in terms of financial resources) RTD programme for robotics research and information society technologies in general[19].

5. Principles of Anthropocentrism on Robotics

The concept to adopt anthropocentric approaches on robotics is very much related to the need of improving the work environment, and to increase the reliability of work procedures in complex and integrated

---

[18] The information can be collected at the European Cordis database (http://cordis.europa.eu/).
[19] In reality, after the ESPRIT designation, the European Commission named this kind of research field the name "Information Society Technologies" or IST.

systems. It is agreed for many years that a better work environment is not merely a physical environment (noise, light, repetitive tasks, etc.). It must include – always – the psychological and social dimensions. These are mostly related with the options for work organization models. When an "intelligent" equipment is introduced to mediate the work relation between people and the material to be transformed, this means that particular care must be given to that equipment, to that technology, to that "intelligence". For such reasons it can be concluded that an approach only based on interfaces improvements is very limited. Social, psychological, ethical dimensions need a veryadvanced research on robotic systems, especially when they are supposed to be integrated as "co-workers" in a manufacturing environment.

In this sense Rauner and Ruth present also an interesting definition: "the concept of anthropocentric production refers to healthy and qualified work, various cooperation and communication options, a maximum of scope for action and shaping on the part of employees (minimization of restrictiveness), technology that is shaped so as to be complementary to human abilities and development potential as well as social and ecological utility of the produced goods (Goods and not Bads)" (Rauner & Ruth 1991: 3). In fact, the concept of anthropocentrism is strongly related with the dimension of working conditions and physical and mental environments.

In the recent decades, the improvement of working conditions has been translated from research results and public debate also to the European legislation. In 1989 the Community Charter of the Fundamental Social Rights of Workers[20] stated that "The completion of the internal market must lead to an improvement in the living and working conditions of workers in the European Community. This process must result from an approximation of these conditions while the improvement is being maintained, as regards in particular the duration and organization of working time and forms of employment other than open-ended contracts, such as fixed-term contracts, part-time working, temporary and seasonal work", and also that "Every worker must enjoy satisfactory health and safety conditions in his working environment. Appropriate measures must be taken in order to achieve further harmonization of conditions in this area while maintaining the improvements made". Such movement for the

---

[20] Published in 9 December 1989.

improvement of working conditions in the manufacturing industries was a basis for the development of anthropocentric experiences with the implementation robotic cells or, in general, with the implementation of Flexible Manufacturing Systems. The basic components of an anthropocentric robotic system can be defined by the following elements:

- Flexible automation, supporting human work and decision-making. It can be considered just a political strategy, but it has clear productivity consequences, as the next elements can demonstrate.
- A decentralized organization of work, with flat hierarchies and a strong delegation of power and responsibilities, especially at shop-floor level. The basic idea to include this element is that this can enable the possibility to react responsively and quickly to a problem
- Reduced division of labour (derived from the previous element)
- Continuous, product-oriented up-skilling of people at work. The need to get involved in the planning, programming and control of production process implies a continued training activity and development of the task competences.
- Product-oriented integration within the broader production processes.

As previously mentioned, an anthropocentric production system can be defined as a coherent set of technological and organizational innovations to improve productivity, quality and flexibility. "The production system that fits this condition is a computer-aided production system strongly based on skilled work and human decision-making combined with leading edge technology. It can be called an anthropocentric production system" (Lehner 1992). In other words, "the strategic goal of anthropocentric shaping of system is to draw man out of his role as a plaything/object of the process and create the prerequisites enabling man to become the subject of production. This means the quality of production work: a) must be qualified and qualifying; b) should raise the level of autonomy of the work/worker and; c) must raise the degree of self-determination of the subject in production. These relatively abstract *characteristics*, however, arose in opposition to Taylorist or technocentric approaches (i.e. mechanistic ones where the production process is seen within the metaphor of the machine with man as a potential disruptive factor and part of the machine) (Rauner & Ruth 1991: 18). Such production systems include normally robotic elements, and research on robotics has been strongly supported by companies and by state institutions for RTD support. That is why such approach deserved (and still deserves) a large attention from the scientific community as from the financial supporters to such research field. But the demands are essentially inter-disciplinary, and not only technological.

With such intentions some focused research institutes develop their activities and research agenda into these topics. An example comes from the US where the Robotics Institute at Carnegie Mellon University is one example. It is the only research institute where it was focused a field on the Anthropocentric Robotics. In that Institute it was agreed the importance of understanding people in order to build better robots. These robots were mostly developed as autonomous systems, and applied to the space research and development. In US some of the most advanced research on user-centred and autonomous systems is done on that sector (space)[21]. In Europe the ESA has been not dealing with such topics. Only the early ESPRIT projects on CIM (Computer Integrated Manufacturing) were during several years the main milestones for such strategies.

Such robotic development can be applied to industrial and human sites, like those where the working conditions are difficult (mining, nuclear power plants, underwater activities). The goals of that project on anthropocentric robotics were to develop a "cognitive model" of how people understand robots, to integrate knowledge about this model into robotic systems, and to evaluate the effectiveness of this integration in improving human-robot interactions. Some studies of human-robot systems were developed under this research programme. One was the study of employees at NASA Ames and their involvement in the Mars Exploration Rover (MER) mission[22]. Another is a study of the interactions among scientists, roboticists, and a semi-autonomous rover as part of the Life in the Atacama project. From their perspective, robots that work with and for people must be

---

[21] Already in 1987 the US National Research Council and NASA held a Symposium on "Human Factors in Automated and Robotic Space Systems".
[22] This study was already completed.

designed not just to adapt to the physical world (the primary emphasis in traditional robotics) but also within the social world [23].

Toward this end, the work in this project has a distinct interdisciplinary character in its blend of the disciplines of design, social psychology, and robotics. This has been one of the most important projects designed in US under the anthropocentric robotics approach. In Europe, the above mentioned project SME Robot uses a similar principle. The approach on the development of specific robotic system for SME (small and medium-sized enterprises) usage has a different content than those that stress mostly the person-machine interfaces

The further development of programming environments is moving continually between the reduction of complexity of increasingly voluminous programs by means of abstract types of data and the increase of the degree of abstraction, which makes the reflection of one's own actions increasingly more difficult. The programming environments are not dealing with a sequence of subsequent steps anymore. All of programming procedures are just in use in more conventional automotive manufacturing companies, although the machine languages and increasingly also the assembler languages are retreating into a few niches. The development with the more important consequences in this area is currently being experienced by the group of professionals in the shape of the establishment of object-oriented programming languages and environments. Not all of them are engineers or computer technicians, but are also skilled operators. Beside a level of abstraction increased yet again, they require a radical "new thinking", that is no longer oriented toward data flows and processes (as in traditional procedural planning) but toward objects and the exchanges between these.

Also in recent years the demand for new and more natural human-machine interfaces has been increasing, and the field of robotics has followed this trend. The speech recognition is seen as one of the most promising interfaces between humans and machines, because it is probably the most natural and intuitive way of communication between humans. For this reason, and given the high demand for more natural and intuitive interfaces, the automatic speech recognition

---

[23] In this same institute another project on People and Robots has been taken and developed in the recent years (http://www.peopleandrobots.org/). It is a group of researchers who are studying how people interact with robots.

(ASR) systems had a great development in the last years. Today, these systems allow, for example, the control of industrial robots in an industrial environment (in the presence of surrounding noise).

Another development is based on the adoption of high-level programming (HLP) techniques can overcome the drawbacks of classical approaches to programming. This can be important to understand how far research in this field is facing challenges and new steps. These types of programming techniques are crucial for the use of industrial robots (and for robot programming in general) since it could help users in the programming tasks, making them easier, especially when they must be applied to robots. The basic idea with HLP systems is to have humans teaching a task solution to a robot using a human-like procedure, which can be done in several ways and at several different levels as already mentioned. This is particularly important in manufacturing environments. And even more important when skilled operators are dealing with robots programming and control and integrated into semi-autonomous working groups in flexible manufacturing systems (FMS) with robots, or simply in robot cells.

The strategic research agenda for robotics presents also aspects related to societal issues. For example, it considers that a more widespread use of robots may lead to further labour displacement and an extension of the digital divide. This may lead to the exclusion of parts of the society from the benefits of advanced robotics. This may seem somehow simplistic once many studies confirmed the non-"technological determinism" and underlined the fact that labour displacement depends on the organizational options and not on the features of the technology itself. On the other hand, and still according to that European research agenda, job profiles can improve as robots take over dangerous, dull and dirty jobs not only in the manufacturing industries. Finally, enhancing the human body through robotics has both positive and negative implications for the able-bodied and disabled. This can be a more recent topic of debate and is presented in several chapters of this book (...) as well in a wide range of studies (Lebedev & Nicolelis 2006; Grunwald 2007; Coenen et al. 2009).

However, more recently, the debate has been developed also over new issues that relate the working perception with autonomous systems (e.g. autonomous robotics). The cognitive task automation, even with visual programming or other user friendly tasks, may lead to over trust, complacency and loss of the necessary work environment situation awareness. This is a major constraint in complex work organizations teamwork, either in service or in manufacturing industries. That can end up into an operational gap, between system developments and its understanding and usability, by operators.

Today one can understand the wider use of the anthropocentrism concept applied to the production architectures, emerging a new value of the intuitive capacities and human knowledge in the optimization and flexibilisation of the manufacturing processes. This includes also the new risk situations that occur with the use of robotic systems. That implies a need to take into consideration qualitative variables in the definition and design of robotic systems, jobs and production systems.

With the development of European research activities (projects, networks, platforms) in the sequence of Framework Programmes of R&D the aims, methodologies, concepts and results changed. If in earlier stages the focus was on the organizational design and on the improvement of working conditions, later the main research focus laid on the software design and integration of new computer science concepts (agents, distribution, object-oriented programming). In the recent years new projects were still based on the development of industrial robotics systems integrating new achievement issued from other related fields of research (service, simulation).

In the manufacturing environment, robotic systems have been used in a wider type of workplaces and it seems that there is 'no general turning away from Taylorism' with all of these experiences on work organization and with alternative organization of automated systems. Indeed, after a period of widespread use of 'lean production concepts' in the early 1990s, the 'pendulum is currently swinging in the opposite direction' whereby many companies are reintroducing more Tayloristic work concepts. The developments of work organisation are very different depending on the specific national, branch (the Scandinavian or the German automotive models, are just examples) and company circumstances and particular market conditions.

The European experiences related to anthropocentric production systems based on the use of skilled workers and flexible technologies adapted to decentralised and participative organisational forms were forgotten and displaced by the so-called "lean production" movement. That anthropocentric production model responds efficiently to the new market demands, but mainly, allows a substantial improvement of the quality of working life (cf. Moniz & Kovács 2000). In fact, the first half of the 90's was strongly influenced by re-engineering (BPR): "to manufacture more and better with less" was the main objective. The rationalisation of operational processes, through the maximum grouping of jobs and tasks, the vertical compression and de-centralisation of decision for an increased flexibilisation, the suppression of wastes, there are the American alternatives to the Japanese challenge. Although a substantial part of re-engineering experiences was not well succeeded, those ideas continued to be largely disseminated (cf. Hammer & Champy 1993).

One obvious point that too often gets neglected is that competitive success based on quality and up-skilling is only one of a number of strategies available to organisations. Others include seeking protected or monopoly markets, growth through take-over and joint venture, shifting operations overseas, cost cutting and the new forms of Taylorism. And all of these have been also achieved with the integration of industrial robotic systems or other integrated automation complexes. Thus, a single trend is not clear.

Once again Rauner and Ruth underline that "a holistic approach to the design of technology and work must involve the consideration of human-centred technical and social criteria from the beginning of the design process. Amongst most contemporary engineering designers, the design of technology [24] and work is still viewed almost solely as a technical concern and it is therefore important that some method whereby human-centred considerations can re-shape this process is made available to designers in order to direct this trend towards anthropocentric principles" (Rauner & Ruth 1991: 20–21). The strong weight of this technology-centred approach is still prevailing in the second decade of the 21st century, against all odds.

6. Technical Systems without Humans or Anthropocentric-Based Systems? Some Concluding Remarks

The actual state of the debate on can be defined when one analyses these projects and networks. In first place, it seems there is a need to relate the working perception with autonomous systems (e.g., autonomous robotics). Such relation did not appear in the decade of the 1990th or even sooner. This is a clear consideration when

---

[24] Including autonomous robotics and systems [ABM].

analyzing research on the new generation of robotic systems in manufacturing. And, second, in the recent years it became also clearer that the cognitive task automation may lead to over trust on technology and technological issues. Although there is a visible need There are very few research projects on social and political issues of anthropocentric strategies in manufacturing. It seems that this can lead to a new problem. The relation between risk, trust and technology development is becoming a clear topic where there is a shortage of studies. As described aboveFthe increase automation tools can lead to an increased complacency and loss of the necessary work situation awareness in highly automated environments.

This trend to over trust autonomous technologies can represent a major constraint in complex work organizations teamwork, where those technologies are applied, ending up into an operational gap, between system developments and its understanding and usability, by operators. In this way, many concepts issued from the work organization analysis, are connected with other concepts such as motivation, alienation, satisfaction, productivity, innovation, flexibility and business processes, learning organizations, networks and virtual enterprises. But these are not tackled by the robotics research. This should be understood as a topic to be researched more in the next future.

In a recent meeting of the EUROP and EURON technology platforms, one official presentation mentioned the "Societal Challenges" of robotics as related to: a) Ageing Society; b) the Climate Change; c) Sustainable Manufacturing, and d) Safety & Security. At the same time, is said that European robotics has much to offer to tackle societal challenge, not only to create awareness, but also through that to improve marketing for robotics. In other words, the robotics technology community, including the equipment manufacturers, understood that the research (and through that, the knowledge) on social dimensions of autonomous systems will also contribute to their marketing aims.

Still connected with those above mentioned dimensions, it is known that 1/4 of European citizens will older than 65 years by early next decade, and twice as many older than 80 years than today (2010). Beside this "ageing society" effect, the climate change will introduce new environmental problems that will affect human health and living conditions. The awareness for a more sustainable manufacturing system is pushing the industry towards new behaviours towards ethics and towards the design of their products and services. Just very few cases can be mentioned [25], but the interest on the need to develop further knowledge of societal issues seems to grow. Slowly, because the counterpart in terms of major support to research on social sciences about these topics in Europe, Japan and US is not yet enough. This means that new specialized areas of robotics (beside Industrial Robotics) are emerging in close relation with new social needs, as the Professional Service Robotics, the Domestic Service Robotics and the Security & Space Robotics. This means that the growing perception of importance of social, political and ethical aspects is revealing also new market niches (that can be of some importance to manufacturers and to innovation support institutions) and new areas for technology development on robotics. At the same time, the development of robotics has contributed to a reduction in the energy consumption in manufacturing processes. This happens because research could develop lighter robots (with new material, and with improved technologies), and also could improve the energy efficiency of robotics. That contributed to improved energy efficiency of manufacturing process due to use of robots, with clear effects in terms of cost reduction. Another implication related with environmental issues, is the possibility that robotics can achieve to reduce material consumption, with less deficient products and efficient use of material (for example, with the painting robots) or low waste production. The previous experiences with anthropocentric systems demonstrated that this implication can be optimized when the development of robots and integrated systems is done together with the involvement and participation of their operators in the shop floor.

Where are the main fields where robotics is still supposed to develop in the next decade? In the recent EUROP meeting they were pointed out:

- Large Structure Manufacturing (incl. civil eng., and at aerospace and shipbuilding)

---

[25] For example, those that were already mentioned when it was referred the case of the European SME Robot initiative in the 6th Framework Programme. Some few more that had experienced the implementation of anthropocentric production systems.

- Robot with Integrated Process Control (self-programming and optimized cycle times)
- Rapidly Adaptable Manufacturing Cell for multi-robots systems
- Coordinated Mobile Manipulator (ceiling mounted robots, wireless control, loop-arrangements)
- Human-like Assembly Robot (flexible two-arm assembly, anthropomorphism)
- Robot Automation for Small-Scale Manufacturing (new robot systems for SME)
- Postproduction Automation (recycling, remanufacturing), with sensor development and for maintenance in under water, dangerous situations, small spaces
- Micro-Manufacturing Robot (for assembling and handling micro-components in multi-stage production lines)
- Robot Assistant in Industrial Environments (maintenance robot, forestry and agriculture robot, de-mining robot, professional cleaning robot, orbital and planetary robot agent and assistant, care robot, surgical robot, rehabilitation robot, logistics robots

Most (if not all) of these fields that need further research have inherent evidence of social and economical impact, and seem to be needed in the near future. Some of these are normally classified as "service robots", and they will probably know an increase of their "population" of (intelligent) machines for the next years. Some of these service robots will be integrated also in the manufacturing sector, as the ones related to maintenance, logistics, inspection and quality control.

The robotic application to under water environment and to detection of fires and catastrophes will be used as fast they can contribute to cost reduction in such operations and have economical evidence of their utility. Here the need for an easy operability and accurate capacity will always be based on human competences, and their development can be only made on the basis of direct collaboration and participation of operators and users.

Health dimensions will be of further interest in the robotics R&D policies. Not only the surgical robots, but also the care giving and rehabilitation robots, and all related to provide missing body elements to handicapped people (legs, hands, arms). This field is perhaps the one where the ethical issues are becoming more decisive to define the bias of technology development. It is possible to experiment highly advanced systems and bionic equipment, but research will encompass the market needs. And these needs are defined by health policies and socio-economical strategies. Either defined by national and regional governments, or by large companies, it remains a governance issue.

The capacities of 'human' intuition and 'human' knowledge must be still a condition for the development of autonomous systems and also conditions for the optimization and flexibilization of manufacturing processes. That would mean alternative options at the organizational level. But, these new organizational qualities associated to the importance of human and social aspects of automation, also include new risk situations that can occur with the wider application of robotic systems.

It is possible that is emerging a new value of the intuitive capacities and human knowledge in the optimization and flexibilization of the manufacturing processes. This would be a pre-condition to understand the human-robot communication needs. If not, there are new risk situations that occur with the use of robotic systems. Finally, for such reason, there is a need to take into consideration qualitative variables in the definition and design of robotic systems, jobs and production systems. Serious research on robotic systems should imply also these issues in order to create good working conditions in future manufacturing.

References

Barata, J. & Matos, L.C. (1993): "Development of a FMS/FAS System: The CRI's Pilot Unit", in: *Proceedings of the European Community-Latin America Workshop on Computer Integrated Manufacturing* (ECLA.CIM'93), Nov., pp. 125–133.

Beck, U. (2000): *Brave New World of Work*, Polity Press, Cambridge.

Bell, D. (1976): *The Coming of Post-industrial Society: A Venture in Social Forecasting*, Basic Books, New York.

Bessant, J. (1989): *Microelectronics and Change at Work*, Geneva, International Labour Organisation.

Brandt, D. (1992): *Advanced Experiences with APS. Concepts, Design, Strategies, Experiences: 30 European Case Studies*, Vol. 2, CEC - FAST, FOP 246.

Braverman, H. (1974): *Labour and monopoly capital*, Monthly Review Press, New York.

Brödner, P. (1990): "Technocentric-anthropocentric approaches: towards skill-based manufacturing", in: Warner, M.; Wobbe, W. & Brödner, P. (eds.): *New Technology and Manufacturing Management*, John Wiley & Sons, Chichester.

Brödner, P. & Latniak, E. (2002): *Sources of Innovation and Competitiveness: National Programmes Supporting the Development of Work Organisation*, Report to DG Employment and Social Affairs, Institute for Work and Technology, Gelsenkirchen.

Bullinger, H.-J.; Korndörfer, V. & Salvendy, G. (1987): "Human Aspects of Robotic Systems", in: Salvendy, G. (ed.): *Handbook of Human Factors*, John Wiley, New York, pp. 1657–1693.

Castillo, J. J. (ed.) (1988): *La automación y el futuro del trabajo. Tecnologías, organización y condiciones de trabajo*, MTSS, Madrid.

Clegg, Ch. & Corbett, M. (1987): "Research and Development into 'Humanizing' Advanced Manufacturing Technology", in: Wall, T. D.; Clegg, Ch. & Kemp, N. J. (eds.): *The Human Side of Advanced Manufacturing Technology*, John Wiley & Sons, Chichester.

Coenen, Ch. (2009): *Human Enhancement*, STOA (report IP/A/STOA/FWC/2005-28/SC35, 41 & 45), European Parliament, Brussels, pp. 220p.

Davis, L. E. & Wacker, G. J. (1982): "Job Design", in: Salvendy, G. (ed.): *Handbook of Industrial Engineering*, John Wiley, New York.

Dore, R. (1973): *British factory, Japanese factory*, University of California Press, Berkeley.

Durand, J.-P.; Stewart, P. & Castillo, J.-J. (eds.) (1998): *Teamwork in the automobile industry. Radical change or passing fashion?*, Macmillan, London.

Ebel, K.-H. (1987): "The impact of industrial robots on the world of work", *Robotics*, Vol. 3, No. 1, March, pp. 65–72.

Emery, F. & Trist, E. L. (1960): "Socio-technical systems", in: Churchman, C. W. & Verhulst, M. (eds.): *Management science. Models and techniques*, Vol. 2, Pergamon, London.

European Foundation for the Improvement of Working and Living Conditions (1992): *First European Survey on the Work*

Environment 1991–1992, Office for Official Publications of the European Communities, Luxembourg.

FAST (1987): *Human work, technology and industrial strategies. Options for Europe, Synthesis of the results of FAST study on "Technology, Work and Employment"*, CEC - FAST, Brussels, November.

FAST (1989): *Human Work in Avanced Technological Environment.* CEC - Fast, June.

Fiedeler, U. & Krings, B. (2006): "Naturalness and Neuronal Implants – Changes in the perception of human beings", *MPRA Paper*, No. 8501, University Library of Munich.

Freyssenet, M. (1995): "La "production réflexive": Une alternative à la "production de masse" et à la "production au plus juste"?", *Sociologie du Travail*, Vol. 3, No. 95, pp. 365–89.

Grunwald, A. (2007): "Converging technologies: Visions, increased contingencies of the *conditio humana*, and search for orientation", *Futures,* Vol. 39, No. 4, pp. 380–392.

Hammer, M. & Champy, J. (1993): *Reengineering the Corporation: A Manifesto for Business Revolution*, Harper Business, New York.

Hertog, J. F. & Schroder, P. (1989): *Social Research for Technological Change: Lessons from national programmes in Europe and North America.* Maastricht: University of Limburg, Maastricht Economic Research Institute on Innovation and Technology (MERIT).

Husband, T. (1992)*:* "Anthropocentric Technologies: The Way Ahead?", in: P. Kidd: *Organisation People and Technology in European Manufacturing*, Official Publications of the European Communities, Luxembourg, pp. 31–48.

Jones, B.; Kovács, I. & Moniz, A. (1988): "Understanding what is the Human Factors in CIM Systems: Some international evidence", in: Rooks, B. W.: *CIM-Europe 1988 Conference: Present Achievements, Future Goals*, IFS Publication, Bedford.

Jürgens, U.; Malsch, T. & Dohse, K. (1993): *Breaking from Taylorism: Changing Forms of Work in the Automobile Industry*, Cambridge University Press, New York.

Kern, H. & Schumann, M. (1984): *Ende der Arbeitsteilung? Rationalisierung in der industriellen Produktion: Bestandaufnahmen Trendbestimmung*, München.

Kidd, P. (1992): *Organisation, people and technology in European manufacturing*, CEC - FAST, Final Report, Brussels.

Kovács, I. & Moniz, A. (1992): "La introducción de sistemas antropocentricos automatizados en Portugal", *Sociología del Trabajo*, No 16, Madrid.

Kovács, I. & Moniz, A. (1994): "Trends for the development of anthropocentric production systems in small less industrialised countries: The case of Portugal", in: Proceedings of European Workshop in Human Centred Systems. Available at: http://mpra.ub.uni-muenchen.de/6551/.

Lebedev, M. & Nicolelis, M. (2006): "Brain–machine interfaces: past, present and future", *Trends in Neurosciences,* Vol. 29, No. 9, pp. 536–546.

Lehner, F. (1992): *Anthropocentric production systems: the European response to advanced manufacturing and globalization.* CEC, Science and Technology Policy, Synthesis Report.

Moniz, A. (1996): *Organizational alternatives for flexible manufacturing systems.* MPRA Paper, No. 6169, University Library of Munich. Available at: http://ideas.repec.org/p/pra/mprapa/6169.html.

Moniz, A. (2007): "The Collaborative Work Concept and the Information Systems Support Perspectives for and from Manufacturing Industry," *Technikfolgenabschätzung – Theorie und Praxis,* Vol. 16, No. 2, Juni, ITAS-FZK, pp. 49–57. Available at: http://ideas.repec.org/p/pra/mprapa/5627.html.

Moniz, A. & Kovács, I. (2000): *Conditions of Inter-Firm Co-Operation in a Virtual Enterprise Concept: The case of automotive sector in Portugal.* MPRA Paper, No. 5658, University Library of Munich. Available at: http://ideas.repec.org/p/pra/mprapa/5658.html

Piore, M. J. & Sabel, Ch. (1984): *The Second Industrial Divide – Possibilities for Prosperity*, Basic Books, New York.

Rauner, F. & Ruth, K. (1991): *The Prospects of Anthropocentric Production Systems: a world comparison of production models*, CEC-FAST, FOP 249, APS Research Papers Series, Bremen.

Ribeiro, L. & Barata, J. (2006): "New Shop Floor Control Approaches for Virtual Enterprises", *Enterprise and Work Innovation Studies*, No. 2, Monte de Caparica, pp. 25–32.

Schraft, R. D. & Meyer, Ch. (2006): "The Need for an Intuitive Teaching Method for Small and Medium Enterprises", in: VDI-Wissensforum et al.: *ISR 2006 - ROBOTIK 2006: Proceedings of the Joint Conference on Robotics. May 15-17, 2006, Munich: Visions are Reality*. Düsseldorf, pp. 10p.

Skordas, Th. & Robrock, K. H. (1997): *An overview of Robotics Technologies in RTD Programmes of the European Community under FP4*. EC, Brussels. Available at: http://cordis.europa.eu/esprit/src/iimrobot.htm.

Valeyre, A. et al. (2009): *Working conditions in the European Union: Work organization I*, Office for Official Publications of the European Communities, Luxembourg.

Warner, M.; Wobbe, W. & Brödner, P. (eds.) (1990): *New Technology and Manufacturing Management*, Chichester: John Willey & Sons.

Zachary, W. & Weiland, M. (1994): "Interface Agents for Effective Human-Computer Coordination in Hybrid Automation Systems", in: Kidd, P. & Karwowski, W. (eds.): *Advances in Agile Manufacturing*, IOS Press, Amsterdam, pp. 313–316.

Ethical and Critical Views on Studies on Robots and Roboethics

Makoto Nakada

Abstract: In this paper, I will try to find out what is behind people's interest in the studies on robots and roboethics in Europe and USA and also try to know why Japanese show very limited interst in these problems. First, in this paper, I will sketch out the tendencies of disucussions on roboethics or 'Ethics and Robotics' in Europe and the USA, focusing on the meaning of 'autonomy' as a fundamental and ambiguous concept in this field. Secondly, I will try to surface the roles of scholars or authors in this field as a 'deleted' factor determining the direction of discussions regarding robots. Finally, I will examine the data of the researches I did in Japan recently in order to know the relation between people's views on robots and their attitudes toward *Seken* as a Japanese life-world. In my view, *Seken*-related views enable people to have a latent ability to criticize the human–robot-interaction, even though they (we) are not aware of this fact clearly.
Keywords: autonomy, roboethics, *Seken*, morality of robots, Theory of mind

1. Introduction: What is Behind the Recent Studies on Robotics and Roboethics?

The main reason I will try to focus on roboethics in this paper arises (at least partially) from my own experiences, i.e., I can't understand why this topic is so eagerly discussed in Europe or the USA. According to Veruggio and Operto, "the name Roboethics was officially proposed during the First International Symposium of Roboethics (Sanremo, Jan/Feb. 2004), and rapidly showed its potential[1]." In fact, so far as I took a look at the related papers or journals, I have to agree with Veruggio and Operto regarding the importance and the need of discussions in this new field. But, on the other hand, whenever I asked the graduate and undergraduate students in my classes studying values and ethics in information society about whether they think roboethics is an important subject for them, most of them answered, 'no.' And in my view, this is the typical attitude

---

[1] According to Veruggio and Operto, "Robotics is a new science still in the defining stage. Public and private professional associations and networks such as IFR International Federation of Robotics, IEEE, Robotics and Automation Society, EUROP -European Robotics Platform, Star Publishing House, have undertaken projects to map the State-of-the-Art in Robotics." (Veruggio & Operto 2006).

towards roboethics in Japan. Some Japanese authors have similar opinions in this regard. For example, Kitano says that Japanese researchers in robotics tend to 'focus on enhancing the mechanical functionality with having little ethical discussion on the use of robots, while in the West, the robotists often discuss the social and ethical problems for applying robots to human societies[2].'

But on the other hand it is true that Japanese culture and society today are full of images of social robots, pet robots, war robots (robots on the battlefields in comics and in movies), Tamagotchi, anthropomorphic machines and zoomorphic robots. And importantly we are clearly under influence of this deluge of images of robots in our society in some visible or invisible ways. In a sense, our unawareness of the importance of robotethics itself might be a subject for ethical discussions.

The following is an excerpt from a press release about PARO showing the results of the research by AIST, the National Institute of Advanced Industrial Science and Technology in Japan. AIST tried to find to what extent 'communication of aged people with seal-type pet robot PARO influences the minds and behaviors of the aged people. According to the report of AIST, this research done in 2003 and 2004 in Japan proved the remarkable potentiality for therapy through communication between PARO and aged people.

> "The effects of robot therapy were evaluated from psychological, physiological and social aspects. (...) In this way, the contact with PARO proved to be therapeutically effective: psychologically, cheering up, exhilarating, and improving the depression; physiologically, remitting stress; and socially, augmenting interaction among the aged and with nursing personnel and bringing bright atmosphere." (AIST 2004)[3]

Another case of technologically mediated communication by a robot produced by AIST in 2009 is a little more confusing. 'HRP-4C' looking like a young Japanese woman is 158 cm in height and 43 kg in weight. AIST designed this robot as a real young woman who shows various facial expressions with a sort of gestures. When HRP-4C made an official debut in front of a lot of news reporters, she (it) showed anger, embarrassment and other artificial emotions. These

---

[2] Kitano 2007; Kitano 2006.
[3] AIST press released on September 17, 2004 (http://www.aist.go.jp/index_en.html).

emotions evoked a laugh on the side of reporters. But we don't know how her female like figure will bring about response to her potential users.

According to an interesting experimental survey done by Mariko Narumi and Michita Imai[4], the artificial voice of robots with a friendly and sympathizing tone can exert an influence on the behavior of the human subjects. The subjects influenced by a robot voice such as "why don't you have a piece of this cake?" ate a piece of cake more frequently than when there was no offering from a robot. Narumi and Imai explain the results of this research in such a way, "We human beings tend to attribute the friendly voice of a machine to the imagined inner minds or emotions of the robot."

This means, in my interpretation, that the artificial voice (and probably, artificial facial expression too) of robots can influence our human minds. Even though we are aware that the voice of a robot (and expression) is artificial or fake, we are likely to be influenced by a machine showing humanlike emotions or feelings. The case of HRP-4C is not an exception. We certainly feel something such as feelings of 'strangeness,' 'friendship,' 'fetishism" in our minds when we see her face and hear her voice. In that case we are under influence of an interaction with this robot in an (at least to a certain extent) unconscious way. Or we might just do 'attribution' in an unintentional way. But in any case we are engaged in contact with robots. And this is a subject we should see from ethical perspectives. In this sense, ethical questions about robots and human-robot-interaction are not confined to people in the West. But in fact these are our own questions too.

2. Autonomy of Robots: Ethical Problems of Roboethics and Robotics

2.1 'Autonomy' as a 'Hot' Topic in the Fields of Robotics and Roboethics

It's my impression that 'autonomy' of robots is one of the most 'hot' topics in the West. A lot of researchers on roboethics or HRI(human-robot-interaction) in Europe and the USA are enthusiastically engaged

---

[4] Narumi & Imai 2003.

in discussions on this topic. On the other hand, we Japanese have difficulty understanding why this topic, 'autonomy of robots' is so important for 'them' in the West. And as we will see later in this section, in not a few cases, 'autonomy of robots' often stands very close to 'responsibility of robot' in the fields of roboethics. At first glance 'responsibility of robot' is an absurd or difficult topic for us Japanese. Of course it is possible for us to understand the discussions on morality of users, designers or manufactures of robots. And we can understand the need of ethical discussions on robots and HRI in this sense. But 'morality' of robots and 'autonomy' of robots are quite different topics for us. Therefore, as a first step, we have to see carefully how various researchers deal with 'autonomy' of robots and 'morality' of robots in their discussions.

John P. Sullins' discussions on 'autonomy' or 'morality' of robots are the typical ones in the West. And we Japanese have difficulty understanding the needs of discussions focusing on morality and responsibility of a robot as a human-made machine. Sullins says, "In certain circumstances robots can be seen as real moral agents. A distinction is made between persons and moral agents such that, it is not necessary for a robot to have personhood in order to be a moral agent[5]." He adds to this view that we can see a robot as a moral agent, on condition that three requirements are fulfilled: 'autonomy,' 'intention' and 'responsibility.' According to Sullins, the requirement of 'autonomy' is achieved when "the robot is significantly autonomous from any programmers or operators of the machine." And the second requirement ('intention') is seen to be achieved when "one can analyze or explain the robot's behavior only by ascribing it to (its) some predisposition or (its) 'intention' to do good or harm." And the third requirement ('responsibility') means that "robot moral agency requires the robot to behave in a way that shows and understanding of responsibility to some other moral agent."

In my view, we have to pay careful attention to the term 'ascription' as well as 'analysis (by human beings)' and 'explaining (by human beings).' These terms are sometimes clearly and sometimes unclearly used in Sullins' paper and consist of almost invisible key terms in his discussions. It seems that Sullins himself makes a clear distinction between 'autonomy,' 'intention,' 'responsibility' (of robots) and 'autonomy, intention and responsibility as a result of ascription by

---

[5] Sullins 2007.

human beings to robots.' But he sometimes doesn't use the term 'ascription (by human beings)' in a clear way. And this lack of the term 'ascription' itself brings about confusion or misunderstanding regarding autonomy and responsibility of robots.

This confusion or misunderstanding sometimes makes us confused, for example, in the case of discussions on 'mind-less morality.' The following is a citation from Floridi's views on 'mind-less morality' by Sullins.

> "If an agent's actions are interactive and adaptive with their surroundings through state changes or programming that is still somewhat independent from the environment the agent finds itself in, then that is sufficient for the entity to have its own agency" (Floridi & Sanders 2004).

We clearly need to add 'to ascribe' or 'can be seen by us or someone (for example, Floridi)' to this sentence when we try to avoid confusion or misunderstanding suggested above.

In this sense, it is useful to cite the views by Rafael Capurro here. He says: "It is, following the Kantian argument, impossible to create an artificial living or non-living moral being because freedom and autonomy are not a quality of sensory natural and/or artificial beings[6]." Sullins or Floridi might just want to ascribe 'certain circumstances' relating some characteristics of robots to robot's autonomous or moral functions. And they might not believe in emergence of an artificial living with autonomous decision or morality. But in any case, the reason they want to do such ascription to robots or 'moral agents' is invisible.

Our critical review on Sullins's paper suggests us robot's autonomy can't be separated from the researchers' own viewpoints. In this context, 'ascription' might be an almost hidden key term. We have to see more carefully various types of discussions or views on 'autonomy of robots' and 'morality of robots' in the West, paying attention on invisible 'ascription.'

---

[6] Rafael Capurro (forthcoming). Towards a Comparative Theory of Agents.

2.2 Various Views on Autonomy of Robots

As I suggested before, one of the most serious problems regarding ethical questions on social robots or human-robot-interaction arises from a simple fact. Very few researchers and authors clearly explain why they need to talk about 'autonomy of robots' or 'morality of robots.' We get an impression that without almost any clear explanation the majority of Western researchers and ethicists are engaged in discussions on robot's autonomy and morality from various viewpoints.

The following is part of Brian R. Duffy's discussions on morality of robots.

> "The issue of moral rights and duties arises from two perspectives. The first is whether a machine should be programmed to be morally capable of assessing its actions within the context of its interaction with people (this includes the evolution of behavioural mechanisms and associated moral 'values')." (Duffy 2006)
>
> "The second perspective is whether it is necessary to have human capabilities in order to be able to assess morality." (Duffy 2006)

At first glance, these discussions on autonomy and morality of robots are interesting. But as in the case of Sullins' paper, the presuppositions of discussions on autonomy and morality of robot are invisible.

Veruggio is one of the famous researchers on roboethics for his 'Roboethics Roadmap.' In one of his papers[7] he starts his discussions on roboethics with simple questions. "Could a robot do 'good' and 'evil'?" "Could robots be dangerous for humankind?" He says, "This paper deals with the birth of Roboethics." But clearly he is apart from the doubt: why he needs to start his discussions on roboethics with such simple questions. "Could robots be dangerous for humankind?" Because of this lack of doubt about his starting-point, his questions on roboethics remain on a superficial level.

In my view, this lack of the origins of questions leads Veruggio's discussions to an unexpected confusion. Veruggio summarized three typical evaluations by the robotics scientists, researchers, and the general public about 'morality' or 'consciousness' of robot. "Robots are nothing but machines." "Robots have ethical dimensions."

---
[7] Veruggio & Operto 2006.

"Robots as moral agents." About the second evaluation, he explains as follows.

> "In this view, an ethical dimension is intrinsic within robots. This derives from a conception according to which technology is not an addition to man but is, in fact, one of the ways in which mankind distinguishes itself from animals. So that, as language, and computers, but even more, humanoids robots are symbolic devices designed by humanity to improve its capacity of reproducing itself, and to act with charity and good" (Galvan, 2003)[8]

This view itself is acceptable. But despite of this clear view, the important point is still unclear: why do studies of robots have ethical dimensions? In this case too, the starting-points of researchers and authors in the filed of roboethics or human-robot-interaction are not visible.

In one of his papers, Peter Asaro tries to combine 'autonomy' and 'morality' of robot with a different and related problem, 'rights (of robot).' He tries to turn our attentions to this problem, saying "how legal theory, or jurisprudence, might be applied to robots?" To avoid making a negative impression on this problem, Asaro adds the following view to his discussions.

> "Most notably, the case of unborn human fetuses, and the case of severely brain damaged and comatose individuals have led to much debate in the United States over their appropriate legal status and rights." (Asaro 2007)

But the presuppositions of discussions on autonomy, morality and legal rights of robot are not clearly explained here as in the other cases.

3. Why Do They Want to Discuss Autonomy of Robots?

I think that now we have to move to the next step and propose our question: why researchers, authors in the field of roboethics, HRI and also robotics want to focus on the problems related to 'autonomy,' 'morality' and 'responsibility'? We have to find out the related invisible motivations or presuppositions regarding these topics. Then

---

[8] This view is presented by Galvan (2003): "On Technoethics" *IEEE-RAS Magazine*, 10 2003/4: 58–63.

we might be able to turn our weak interest in these topics into a different direction. In this session I will do an effort to let the almost sunken and unclear motivations or presuppositions rise to the surface.

3.1 Decision by Robots and Standing Point of Researchers

The strong interest in the topics regarding robot's autonomy and responsibility as well as the related issues in Europe and the USA can't be separated from the motivations of the researchers and users of robots. This is an impression we can get when we read the following news by AFP, which appeared on August 12, 2009. This news article suggests us that there is a need of development of 'autonomous' robots for certain potential users.

> "Going off to war has always meant risking your life, but a wave of robotic weaponry may be changing that centuries-old truth. The "pilots" who fly US armed drones over Afghanistan, Iraq and Pakistan sit with a joystick thousands of miles (kilometers) away, able to pull the trigger without being exposed to danger. Other robots under development could soon ferry supplies on dangerous routes and fire at enemy tanks. The explosion in unmanned vehicles offers the seductive possibility of a country waging war without having to put its own soldiers or civilians in the line of fire." (Dan De Luce (AFP) 2009)[9]

We can easily imagine that the military leaders want to develop some sort of technologically advanced robots. If these robots can decide to kill or 'terminate' the 'enemies' by themselves, "the tantalizing scenario of 'pain-free' military action[10]" might come true. In this case, the robots working automatically or 'autonomously' are independent of the operators or of 'real' soldiers. The following is the remaining part of AFP news about "Robots at war."

> "US officials insist a human will always be "in the loop" when it comes to pulling the trigger, but analysts warn that supervising robotic systems could become complicated as the technology

---

[9] "Robots at war: will humans stay in the loop?" by Dan De Luce (AFP) (12.08.2009).
(http://www.google.com/hostednews/afp/article/ALeqM5hn18cylh7aHOhpNClP7eBfOD6E4g).
[10] "Robots at war: will humans stay in the loop?" by Dan De Luce (AFP) – Aug 12, 2009.

progresses. Military research is already moving toward more autonomous robots that will require less and less guidance."

Clearly, this need urges people toward another need of ethical and political discussions on 'autonomy' of robots at the same time. In order to use military robots efficiently on the battlefields, the military leaders and the political leaders have to consider carefully how to avoid the moral problems. If military robots violate war rules, the human leaders will receive criticism. They have to avoid this criticism. Robot soldiers as well as human soldiers must obey Geneva Convention. They can't use their lethal weapons 'at will.' As we saw before, in the discussions on ethical robots, discussions on 'autonomy' and discussions on 'morality' as well as 'responsibility' are often 'coming together.' We can see this 'coming together' in the case of military robots in a typical way. The following article on military robots shows us a typical case of 'coming together' of 'autonomy,' 'morality,' 'expectation' and 'anxiety' regarding robot.

> "Governments around the world are rushing to develop military robots capable of killing autonomously without considering the legal and moral implications, warns a leading roboticist. But another robotics expert argues that robotic soldiers could perhaps be made more ethical than human ones. Noel Sharkey of Sheffield University, UK, says he became "really scared" after researching plans outlined by the US and other nations to roboticise their military forces. He will outline his concerns at a one-day conference in London, UK, on Wednesday. Over 4000 semi-autonomous robots are already deployed by the US in Iraq, says Sharkey, and other countries - including several European nations, Canada, South Korea, South Africa, Singapore and Israel - are developing similar technologies."[11]

At first glance, it sounds quite strange to see the researchers or authors discuss various cases of war rule violation by robot from ethical perspectives. Because we know that autonomous decisions by robot is just a plan at present. We have an impression: these are just pedantic imagination by robot enthusiasts. But in my view, this means certain sorts of 'a m o c k trial,' 'judicial precedents' or 'simulation of war rule violation by robots.' And in this sense, ethical discussions on

---

[11] " 'Robot arms race' underway, expert warns," *New Scientist*, 27 February 2008. (http://www.newscientist.com/).

'autonomy' or 'morality' of robots by researchers are a sort of simulations combined with hope and anxiety as the following discussions suggest.

> "Clearly, there are fundamental ethical implications in allowing full autonomy for these robots. Among the questions to be asked are: Will autonomous robots be able to follow established guidelines of the Laws of War and Rules of Engagement, as specified in the Geneva Conventions?; Will robots know the difference between military and civilian personnel?; Will they recognize a wounded soldier and refrain from shooting?" (Lin et al. 2008)[12]

3.2 Disappearance of Personal Decisions, Morality and Consciousness

We can see a change in autonomy, independency or morality on the side of human beings leading to ethical discussions on robots. Patrick Lin, whose report to US Navy was quoted above, talked about these situations in an interview with a reporter of *Times*.

> "There is a common misconception that robots will do only what we have programmed them to do," Patrick Lin, the chief compiler of the report, said. (...) The reality, Dr Lin said, was that modern programs included millions of lines of code and were written by teams of programmers, none of whom knew the entire program: accordingly, no individual could accurately predict how the various portions of large programs would interact without extensive testing in the field – (...)."[13]

Lin's solution quoted below sounds strange. He doesn't doubt the need of the fighting robot. But at the same time Lin's discussions suggest at least indirectly human's changes in today's society characterized by division of work, networking. This is what we have to take into consideration.

---

[12] Patrick Lin, George Bekey and Keith Abney (2008). *Autonomous Military Robotics: Risk, Ethics, and Design*. Prepared on: December 20, 2008 (*This work is sponsored by the Department of the Navy, Office of Naval Research,under award # N00014 - 07 - 1 - 1152*). This report is part of official studies on ethical military robots by US Navy.

[13] "Military's killer robots must learn warrior code," The Times(on-line), February 16, 2009.
(http://technology.timesonline.co.uk/tol/news/tech_and_web/article5741334.ece)

> "The solution, he [Patrick Lin] suggests, is to mix rules-based programming with a period of "learning" the rights and wrongs of warfare."[14]

These situations are the same in the case of so-called networked robots ('Network Robot Systems'). According to Sanfeliu, Hagita and Saffiotti[15],'Network Robot Systems' (NRS) include(s) the following elements: physical embodiment; autonomous capabilities; network-based cooperation; environment sensors and actuators; human-robot interaction. We can easily imagine that within these systems the needs of autonomous humans well as autonomous robots are decreasing; because systems themselves have their own 'eyes' and 'ears' and actuators. In these cases, this complex combination of elements means disappearing of the independent agents. And, clearly, even the concepts of 'autonomy' and 'morality' are getting meaningless; because independent agents don't exist any more. It is difficult to imagine 'autonomy without agent.' The development of the highly organized and networked systems such as NRS might foretell the end of need of discussions on 'autonomy of robots.'

3.3 Abuse of Theory of Mind

As some psychologists and roboticists suggest, 'Theory of mind(ToM),' 'recursive ToM,' or 'recursive mental states' might provide us with good hints on our strange experiences about human-like robots: we cannot but have a feeling that robots have something like 'minds' within their mechanical bodies. But at the same time, 'Theory of mind' often leads to forgetting of the mediating role of researchers standing between robots and us humans. As in the case of autonomous robot, the researchers work as intermediator between robot and humans, trying to combine mechanical minds and human minds through interpretation of a psychological thoery.

The following are a list of explanations on 'ToM' or 'recursive ToM.'

> "Many authors have assumed that the proximate mechanism which facilitates this highly developed sociality is Theory of Mind (ToM), which is defined as the ability to make inferences about the beliefs and desires of other people." (Liddle & Nettel 2006)

---

[14] "Military's killer robots must learn warrior code," The Times(on-line).
[15] Sanfeliu, Hagita & Saffiotti 2008.

> "Researchers working on adults, however, have developed more demanding tasks by probing recursive ToM understanding (second level: inferences about a belief about a belief; third level: inferences about a belief about a belief about a belief, and so on, up to fifth or in one case, eighth, level." (Liddle & Nettel 2006)

> "Known as theory of mind, the ability to infer another's perspective - emotional, intellectual, or visual - improves with age. (...) In a typical test, kids watch two puppets - Sally and Anne - play with a marble, then put the marble back in a box. Anne "leaves" and Sally grabs the marble, plays with it, and then returns the marble instead to a bag. Where will Anne first search for the marble, researchers ask the children. "Before four, kids say she's going to look in the bag, but after four they know she has a false belief," says Iroise Dumontheil, a cognitive neuroscientist at University College London, UK, who led the new study." (Callaway 2009)[16]

Clearly, in this case, the roles of researchers are important as in the case of 'autonomy of robots.' The researchers play a role of judge who decides whether we should accept certain theories or perspectives or not.

Therefore if the researchers and authors have no critical or ethical eyes toward use of 'Theory of mind' and toward their invisible roles as 'judge', this lack of critical and ethical eyes might bring about another ethical problem. As the report of *New Scientist* indirectly suggests, 'Theory of mind' sometimes only refers to the 'recursive mind states': inferences about a belief about a belief. In fact, the children who can't recursively infer that 'Anne' has a false belief are never 'mindless.' They have minds, but their function of minds related to 'Theory of mind' is comparatively weaker. In this sense, the following discussions sound strange. Because these discussions on robot's mind are apart from human observers who judge whether this is the case of 'Theory of mind' or not.

> "A robotic system that possessed a theory of mind would allow for social interactions between the robot and humans that have previously not been possible. The robot would be capable of learning from an observer using normal social signals in the same

---

[16] "Why teenagers can't see your point of view," *New Scientist* (05 February 2009 by Ewen Callaway).
(http://www.newscientist.com/article/dn16535-why-teenagers-cant-see-your-point-of-view.html).

way that human infants learn; no specialized training of the observer would be necessary. The robot would also be capable of expressing its internal state (emotions, desires, goals, etc.) through social interactions without relying upon an artificial vocabulary." (Scassellati 2002).

A similar and more serious criticism might be made against neglect by the researchers and authors, when we consider individual differences in ToM performance. As the following remarks suggest, levels of 'Theory of mind' might bring about an unexpected ethical problem.

"It is likely that individual differences in ToM performance persist through school years and beyond, into adulthood. However, a problem with the assessment of such variation is that, with the exception of individuals with pervasive disorders such as autism, performance on first-order belief tasks is usually at ceiling in individuals over 5 years old." (Liddle & Nettle 2006)

Take the interaction of human children and human-like robot Wakamaru, for instance. The following is cited from Wakamaru's formal website.

"The start of a new relationship between humans and robots. 'Wakamaru' is a real robot created to live with humans. It will bring fun and friendship into your lives and feel just like family."[17]

If we see the closed relations between Wakamaru and children, we might not find out any serious problems here. But actually, these closed relations between robots and children would be impossible, if human adults such as parents of children, producers of the robots, observers of HRI, researchers of HRI are not there as hidden participants. In this sense, the closed-bilateral-relations between a human child and Wakamaru is really a trilateral relation. And as a result of this invisible trilateral relation among a child, Wakamaru, the adult third person, an unexpected confusion regarding levels of recursive ToM might happen there. The child without developed ToM might believe that Wakamaru is a living thing and a real friend. Wakamaru might artificially interact with the child, following the programs based on ToM. The child himself can't understand ToM because his/her mind doesn't have any ToM. The third person might

---

[17] http://www.mhi.co.jp/kobe/wakamaru/english/live/index02.html.

ascribe the interaction between the child and Wakamaru with artificial ToM installed into it to Wakamaru's mind. If the observer as the third person is not aware of his privileged position, this is a serious ethical problem. The observer is the only person there who can see what is happening there.

4. Japanese Various Perspectives on Life in this World and their Ethical Views on HRI

Through our ethical/critical examination of trends and tendencies of ethical studies on robots, we could see that the ethical positions of researchers, authors and scientists themselves consist of the ethical problems of robots in various cases. And we could see that some of the ethical problems on robots, for example 'autonomy,' can be blurring without careful attention on the invisible positions of researchers or authors. In many cases, the positions of researchers and authors of roboethics are invisible. But these invisible or 'deleted' positions can determine what sort of ethical or unexpected problems will emerge. To put this another way, we can say that robots exist in three different modes of temporality: the present, the future and the past. The roboticists design robots at the present moment to fulfill the tasks in the future. When robots work at the present moment, this working is predetermined by the simulation of the past time. This unconscious or invisible linkage of time is made possible only by human beings who can live in different time modes. But if they forget this ontological meaning of time, i.e., time depending on human existence, the questions about robots would shrink into mere mechanical ones. In this case, robots with 'autonomy' are seen just as entities.

In this sense, the following remarks by Rafael Capurro, which are based on Heidegger's discussions are important in that they enable us to be aware of distinctions between ontic views on robots and ontological views on robots.

> "One of Heidegger's main discoveries with regard to the question of Being was that for metaphysics the sense of Being is presence. As presence – together with past and future – is a dimension of time it follows that time is the hidden horizon of the metaphysical interpretation of Being. According to Heidegger metaphysics "forgets" temporality in its full three dimensionality by holding only to the one-dimensional sense of presence or "standing presence-at-

> hand." If the digital casting of Being by holding only to the one-dimensional sense of presence forgets the question of Being in its full three dimensionality it "changes over" into digital metaphysics." (Capurro 2006)

Capurro refers to the difference between 'ontological views' and 'ontic views' regarding information technology as well robotics. In the case of discussions on robotics and roboethics, we can understand as follows. 'Ontological views' are related to human existence (hopes, anxiety, imagination, presuppositions, criticism, modes of temporality and so). And 'ontological views' on human-robot-interaction are also realted to this sort of human existence. 'Ontic views' are apart from this sort of human existence. The meanings of robots, in this case, are separated from human interpretation grounded on human existence.

 In my own interpretation, these 'ontological' remarks by Capurro and Heidegger provide us with critical/ethical viewpoints on robots. From these viewpoints we can see that some researchers and authors often fail to notice the differences between ontic views and ontological views and also fail to notice their own ways of life behind the discussions on robots. They might think that their discussions are purely technological and theoretical. But actually these discussions are not apart from their standing-points as well as their existence related to modes of time. To put this another way, in the case of ethical discussions on robots in the West, separation of ontic views from ontological or existential views often arises. And this separation causes invisible but serious confusion. This confusion might worsen when another separation on the ontic level happens. Even on the ontic level, clearly, the concept of autonomy of robot can't be separated from other related concepts such as symbolism, connectionism, oscillation and entrainment. In this sense, we can see that both on ontic and ontological level 'autonomy (of robot) ' is dealt with as an isolated concept. But actually, 'autonomy (of robot)' can't be apart from our understanding of meanings of life such as "what is good life for elderly people with use of robot for their care?" "what is pain-less war on the battlefields?" or "what is the difference between beings and Being?"

I think that through these critical and ethical examinations on roboethics or HRI we are now close to our own standing-point. From this standing-point, which we wanted to get in the beginning of this paper, we can see the following points more clearly than before. Why the 'Western' researchers and authors in the fields of roboethics are so eagerly engaged in discussions on 'autonomy,' 'morality' and 'responsibility' of robots? How we can evaluate their ethical discussions? We can say regarding these points as follows. They are right in a way in that robots or robotics can't be separated from the ethical questions on the meanings of robots and human-robot-interaction. But on the other hand, they fail to notice that their own existence is part of those ethical problems.

The third questions in this paper, i.e., Japanese views on ethical issues on robots seem to be able to be explained likewise: we are influenced by combination of ontic views and ontological views incorporated into robots like PARO, HRP-4C, Wakamaru but we are not aware of this 'coming together' of ontic views and ontological views behind PARO's cuteness, HRP-4C's smile, Wakamaru's friendliness.

In the rest of this paper we are going to deal with the invisible 'coming together' of human beings and robots in Japan. In my view, Japanese outlooks on robots and human-robot-interaction reflect 'coming together' of ontic views and ontological views as well as 'coming together' of two aspects of life for Japanese today.

4.1 Japanese Robots in *Seken* as a Japanese Critical Life World or a *Ba*(Place) for Virtue and Vice

Robots and human beings or ontic views and ontological views 'come together' in an invisible way in 'Western' cultural-social contexts as well as in Japanese cultural-social contexts. In the last part of this paper, I will focus on some cases of 'coming together' of Japanese people and robots by using our research data of the survey. I and my colleagues conducted this survey in 2008 in Japan to find people's attitudes toward their life in today's world. In my interpretation, Japanese people's attitudes toward their life and their views on ways of life are characterized by combination of a traditional cultural aspect and a modernized aspect. We can call the traditional cultural aspect '*Seken*' or '*Seken*-related *Ba*(place)' and we can call the modern and highly informatized aspect '*Shakai*' or '*Shakai*-related *Ba*(place).' The term *Seken* derives from *Se*(life) and *Ken*(in-between or locus), i.e., *Ba* between I and you or *Ba* for (good or bad) life. In *Seken*, people always evaluate life from perspectives leading to shared meanings of good-bad human relations and to virtue and vice. *Shakai* is a translated term from modern Western languages (English, French, German) into Japanese. We imported *Shakai*, originally Society, and the related concepts from Western countries in Meiji Era (1868-1912). In *Shakai*, the concepts and values imported from the West such as democracy, human light, individualism and rationality are important.

As we will see later in this section, Japanese views on robots and HRI proved to be part of *Seken*-related meanings or values. Clearly, robots are based on modern technology and belong to *Shakai* through such concepts as autonomy, development (of technology) or independency. But at the same time robots belong to *Seken* in various ways. In this sense, we can see Japanese robots as combination of *Seken* and *Shakai* as well as combination of ontic views and ontological views. Therefore, we have to see our Japanese *Seken* in order to know our Japanese robots. In the following section, we will mainly deal with relation between Japanese *Seken* and Japanese robots.

4.2 Japanese Views on *Seken*

The following figures show Japanese views on *Seken* or *Seken*-related views measured by the research questionnaires of a set of researches done by I and my co-researchers in the past 15 years.

What I tried to point out in the previous papers (see: notes 18) is that Japanese people of modern era remain in a traditional, indigenous and non-rational aspect of the world called *Seken*, although they (we) live doubtlessly in a more modern, rationalized and informatized sphere of the world called *Shakai* simultaneously. According to my own interpretation, people's views on various problems reflect these two aspects or spheres. The important thing here is that we can't neglect the strong influence from *Seken*. This might be surprising for Western readers as well as for Japanese readers who don't pay attention on influence from *Seken*. But as our research data show, *Seken* plays a crucial role in Japan today as a sort of horizon on which various opinions, attitudes, feelings, thinking from different sources of life, history, memories, experiences can 'come together.'

Fundamentally, *Seken*[18] is an aspect in which shared values and meanings including morals, social ethics, criteria of value judgments, common senses, desirable aims of life, beliefs in respectable behaviors, frameworks of understanding life, virtues and the like. As I

---

[18] In regard to *Seken*, see: Nakada 1982, Nakada & Tamura 2005, Nakada 2006.

suggested somewhere else (see: references referred to in the note 18), one of the most remarkable characteristics of *Seken*-related meanings is that they still exist in the minds of Japanese today, although *Seken*-related meanings derive from our past cultural, political, historical, and religious experiences related to Buddhism, Confucianism, Shinto, Bushido (moral and ethics of Samurai), traditional views on nature, orientation to solidarity and so on.

As the following tables show, the majority of Japanese regard various *Seken*-related views as something worthy of respect in their everyday life. (Concerning the researches shown in the following tables, see the references in Note 26 and Note 27[19].)

Table 1. Sympathy with *Seken*-related meanings in Japan

	1995 G	2000 G	2002 G	2003 G	2006 G	2007 G	2008 G
Distance from nature	73.6%	-	82.6	79.0	78.2	-	79.8
Honest poverty	83.7	81.5	84.4	80.3	83.2	-	84.0
Destiny	84.4	79.0	77.9	76.0	81.6	-	81.2
Denial of natural science	88.5	88.3	90.7	88.7	89.6	-	86.2
Criticism of selfishness	85.5	88.3	90.0	90.3	84.2	-	90.2
Powerlessness	71.9	64.8	69.2	62.0	60.8	-	73.4
Superficial	73.3	65.6	70.8	62.7	60.4	-	71.0

---

[19] '1995G' and the other researches were conducted by the author and his colleagues. '1995 G'= research conducted in Tokyo in 1995 (587 respondents collected through random sampling of over 20 years old). '2000 G'= research done in Tokyo metropolitan area in 2000 (611 respondents collected through random sampling of over 20 years old). '2002 G'=research conducted in 2002 in Japan to a range of 25- to 44-year-old survey monitors (569 respondents) selected by a research company in Japan. This survey was designed as quota sampling, and ratios of gender and age were quoted from the 2001 World Internet Project Japan (WIPJ) report. '2003 G'=research conducted in a similar method to 2002G (876 respondents). '2006G' was done in 2006 similarly to 2002G (500 respondents of 20-49 years old). 2007G (1200 respondents of 25-49 years old) and 2008G (500 respondents of 25-44 years old) were done in 2007 and 2008 similarly to 2002G.

cheerfulness							
Belief in kindness	-	68.1	73.1	71.5	76.4	84.3	77.2
Scourge of heaven	62.7	49.5	-	-	55.0	-	-
Warnings form heaven	-	-	-	-	-	63.1	67.4

1) Table 1 shows the percentages of the respondents who said "agree" or "somewhat agree" to *Seken*-related statements. These statements are such as: "Within our modern lifestyles, people have become too distant from nature" (Distance from nature); "People will become corrupt if they become too rich"(Honest poverty); "People have a certain destiny, no matter what form it takes"(Destiny); "In our world, there are a number of things that cannot be explained by science"(Denial of natural science); "There are too many people in developed countries (or Japan) today who are concerned only with themselves" (Criticism of selfishness); "In today's world, people are helpless if they are (individually) themselves" (Powerlessness); "In today's world, what seems cheerful and enjoyable is really only superficial" (Superficial cheerfulness); "Doing your best for other people is good for you" (Belief in kindness); "The frequent occurrence of natural disasters is due to scourge of heaven" (Scourge of heaven); "Occurrences of huge and disastrous natural disasters can be interpreted as warnings of heaven to people" (Warnings from heaven).

It might be quite strange at first glance (at least for the 'Western' people and also for the Japanese who are aware of only *Shakai* and not *Seken*) to see that majority of Japanese people today have these views in their minds because almost everyone knows that Japan of today is an industrialized and modernized country. But as our researches in the past 10 or more than 10 years show, it is clear that Japanese people still live in an aspect of the world or life-world, which might be called *Seken*.

Originally, *Seken* consists of two different meanings, *Se* and *Ken*. *Se* means this world and *Ken* means 'between' or '*Ba*(place).' So *Seken* means 'the world as *Ba* for people motivated by good human relations or by orientation to virtuous (or sometimes vicious) life.' In my (the author's) own view, this '*Ken*' or '*Ba*' is an imaginative locus where different meanings of objects and events 'come together.' In fact, as our research findings suggest, people's interpretation of robots and their ways of life 'come together' within *Seken*.

4.3 Japanese Views on Robots and HRI

The following is the result of the research (2008G) I conducted in 2008 in Japan. The following figures show the respondents' views on robots or HRI.

Table 2. Views on robots in Japan (Data: 2008G)(N=500)(What are your thoughts about various views on robots shown in the following list?)

	Agree	Somewhat agree	Neither agree nor disagree	Somewhat disagree	Disagree
1. To leave handicapped or elderly persons in the care of robots worsens isolation of them from societies even though this idea seems to be appropriate at first glance.	7.2%	35.0	42.2	12.8	2.8
2. It is very natural when children show sympathy or some kind of affection towards virtual creatures like Tamagotchi.	2.4	31.4	39.4	20.0	6.8
3. It is good for children to know the meanings of life through their taking	0.8	24.4	41.4	23.4	10.0

care of virtual creatures like Tamagotchi.					
4. Robots should be given similar rights in the future as fetuses or patients in a coma without consciousness or awareness.	0.8	8.6	44.6	25.8	20.2
5. Robots are expected to be a subject of affection or consideration in the future just as the earth, mountains, rivers are treated so, even though they have no life.	1.4	20.0	50.2	18.0	10.4
6. To leave children in the care of robots would be better than to leave them alone without any care.	0.6	19.0	41.4	24.0	15.0
7. To provide robots with capability of expression of their emotions such as pains	4.0	27.4	43.4	15.2	10.0

would be good in order to prevent (avoid) cruelty or maltreatment to them.					
8. It is natural for some people to get mad when their avatars are insulted, because they feel that the avatars are part of themselves.	3.0	22.8	47.8	17.0	9.4
9. Most of people would avoid maltreating robots if the robots are created to be very similar to us.	0.6	11.8	55.0	20.2	12.4
10. Friendly robots like pet robots for the purpose of human-robot communication are just fake because they have no real minds or feelings.	6.0	16.6	56.4	15.6	5.4
11. It would be very good to use robots as	2.0	34.4	43.2	14.2	6.2

the purpose of education for children at schools in order to promote effects of education.					
12. To use robots on the battlefields would be good because we can reduce the number of casualties of warfare.	4.2	10.4	40.8	19.6	25.0
13. To use robots to do domestic chores would be good because we can lesson the burdens of family members.	7.2	33.0	39.2	15.4	5.2
14. It would be very good to develop new robots, which have a lot of knowledge and are also human-friendly for use as shop assistants.	5.8	25.4	47.8	13.8	7.2

We can say that Japanese views on robots are not clear in many aspects (as we discussed above). But when we conduct factor analysis on the data of this table, we can get four factors as follows. We can understand that this (latent) factors suggest us an important thing: 'coming together' of robots and humans already emerges, even if people are not clearly aware of this 'coming together.' In fact, we can see this 'coming together' in many ways as the following findings show.

Table 3. Factor Analysis (principal factor analysis, Varimax rotation) for 'various views on robots' (Data: 2008G)

Factors	Contributing Values and Factor Loading
Interest in usefulness of robots	13. To use robots to do domestic chores would be good because we can lesson the burdens of family members. (.834) 14. It would be very good to develop new robots, which have a lot of knowledge and are also human-friendly for use as shop assistants. (.796) 11. It would be very good to use robots as the purpose of education for children at schools in order to promote effects of education. (.585) 6. To leave children in the care of robots would be better than to leave them alone without any care. (.511)
Interest in emotional interaction of robots and humans	4. Robots should be given similar rights in the future as fetuses or patients in a coma without consciousness or awareness. (.720) 5. Robots are expected to be a subject of affection or consideration in the future just as the earth, mountains, rivers are treated so, even though they have no life. (.687) 7. To provide robots with capability of expression of their emotions such as pains would be good in order to prevent (avoid) cruelty or maltreatment to them. (.659) 8. It is natural for some people to get mad when their avatars are insulted, because they feel that the avatars are part of themselves. (.520)
Interest in robots as virtual	3. It is good for children to know the meanings of life through their taking care of virtual creatures like Tamagotchi. (.846) 2. It is very natural when children show sympathy or some

creatures	kind of affection towards virtual creatures like Tamagotchi.(.617)
Criticism of use of robots	1.To leave handicapped or elderly persons in the care of robots worsens isolation of them from societies even though this idea seems to be appropriate at first glance. (.551) 10. Friendly robots like pet robots for the purpose of human-robot communication are just fake because they have no real minds or feelings. (.434)

('Contributing value(s)' mean(s) that each factor consists of these items (views).)

The following table (Table 4) shows the correlations between Robot factors and *Seken*-related views. Through our analysis on *Seken*-related meanings based on our previous researches, we could find out the fact that *Seken*-related meanings are related to criticism of various social and political problems. In fact, in the case of 2008G as in the other cases of our researches, we could find the same tendencies regarding *Seken*-related meanings. The (fairly) strong relations between one of Robot factors, 'Criticism of use of Robots' and various *Seken*-related meanings reflect, I think, these critical views included within *Seken*-related meanings.

Table 4. Relations between Robot factors and *Seken*-related meanings (2008G)

	Interest in usefulness of robots	Interest in emotional interaction of robots and humans	Interest in robots as virtual creatures	Criticism of use of robots
Distance from nature	ns	ns	ns	.249**
Honest poverty	.124**	ns	ns	.227**
Destiny	ns	.119**	ns	.192**
Denial of natural science	ns	ns	ns	.263**
Criticism of selfishness	.172**	ns	ns	.287**
Powerlessness	.121**	ns	ns	.161**

Belief in kindness	.093*	ns	ns	.343**
Warnings form heaven	ns	.132**	ns	.132**

1)**=p<0.01, *=p<0.05, ns= non (statistically) significant

The figures in these tables might provide us with a variety of hints with which we can interpret our relations with robots. One of the interpretation we can get from these figures is the finding that human-robot relations reflect human attitudes toward meanings of life, this world, human relations in this life.

4.4 *Seken*, Japanese Imagination and Robots

What we are tying in this section is to surface certain hidden values, views, frameworks of thinking and ways of feeling behind Japanese views on robots. And through our examination of our research data, we could succeed in surfacing the finding that Japanese views on robots might be considered to reflect Japanese views on life, this world, human relations in this world called *Seken*. The following table (Table 5) adds another finding to the results of this examination: critical or ethical views on robots are intertwined with Japanese ways of feeling, evaluating, imagination. To put this another way, Japanese views on robots are based on Japanese life today. And their (our) life reflects or influences a set of ethical, imaginative and critical views on this world.

Table 5 shows the correlations between 'robot factors' and 'Japanese views on imaginative experiences.' It is interesting to interpret these findings shown in this table, because we can easily imagine the location of Japanese robots in Japanese culture. So far as this table shows, Japanese interest in robots, their expectation, their fear regarding HRI, and even their criticism of robots are strongly intertwined with their imaginative experiences in their daily life, particularly their life related to media use.

Table 5. Correlations between 'robot factors' and 'Japanese imagination' (Data: 2008G) (N=500)

	I sometimes feel that places of location or places of making a film appear 'shining' or 'beaming with attractiveness,' even though they are familiar and original places in my daily life. (Agree or Somewhat Agree =58.2%)	I am sometimes influenced by imaginative characters in films or TV dramas with whom I feel strong sympathy, even though I know that they are not real persons. (Agree or Somewhat Agree =42.0%)	I sometimes share a deep sense of emptiness with characters in films or TV drams with whom I feel strong sympathy when they die or fail in life. (Agree or Somewhat Agree =28.8%)	I sometimes have a sense of frustration when I know I can't do anything to help the characters worthy of strong sympathy of films or TV dramas with solving their worries. (Agree or Somewhat Agree =28.2%)
Interest in emotional interaction of robots and humans	.109*	.281**	.301**	.330**
Criticism of use of robots	.225**	.228**	.174**	ns

1)**=p<0.01, *=p<0.05, ns= non (statistically) significant
2) Percentages within brackets show added percentage of 'agree' and 'somewhat agree' of respondents to each views on imaginative experiences.

In my own view, this complexity of views on robots of Japanese people reflects invisible strata or *Ba* of Japanese life. These strata or *Ba* are likely to include *Seken*-related meanings, frameworks of thinking deriving from *Shakai*, ontic views on technology and ontological views on human-machine or human-human relations. Therefore, complexity of Japanese views on robots at least partly arises from this invisible strata or *Ba*. In this sense, in order to analyze our views on robots and to develop our own field for roboethics, we have to know the structures of this invisible strata or *Ba*.

5. Concluding Remarks

I started the discussions in this paper with the questions, 'Why are the studies of roboethics or HRI important for "Western" researchers and why are the necessities of these studies difficult for us Japanese to understand?' Although the discussions on these questions here are based on rather limited material, but on the other hand, clearly, we are very close to our answers through these discussions. In my view, these answers or conclusions might be considered to be a new step towards the (possible) new development in the fields of roboethics or HRI for us Japanese and probably for (some) 'Western' researchers. When we can see the 'deleted' intentions or positions of researchers or authors of roboethics or HRI, we are very close to, I believe, the hidden 'presuppositions' or 'problems' of studies in the fields of roboethics or HRI such as 'separation of ontic views on robots from ontological views on robots,' 'unawareness of different modes of temporality,' 'deletion of "ascription",' and so on. And as we suggested above, we have to pay our attention on dualism of Japanese life reflecting two loci, *Seken* and *Shakai*. But as I said before, the discussions on these questions here are based on limited material, so when we want to move towards the new stage, we have to see more different cases of discussions in these fields, including alternative views on roboethics and robotics grounded on ontological perspectives as well as on recent researches about robotics and artificial intelligence. And concerning our tasks in the future discussions, we have to see similar or different cases of other cultures in the 'East' too. Our discussions on robots are usually confined to limited cases reflecting Western views and partly Japanese views.

References

Asaro, P. M. (2007): "Robots and Responsibility from a Legal Perspective", a paper for *ICRA'07 2007 IEEE International Conference on Robotics and Automation*, 10-14 April 2007, Roma, Italy (Full Day Workshop on Roboethics Rome, 14 April 2007).

Capurro, R. (forthcoming): "Towards a Comparative Theory of Agents", *Proceedings of the panel on Autonomic Computing, Human Identity and Legal Subjectivity hosted by Mireille Hildebrandt and Antoinette Rouvroy*.

Capurro, R. (2006): "Towards An Ontological Foundation of Information Ethics", *Ethics and InformationTechnology*, Vol. 8, Nr. 4, pp. 175–186.

Duffy, B. R. (2006): "Fundamental Issues in Social Robotics", *IRIE* 2006, vol. 6 (Ethics in robotics), pp. 31–36.

Floridi, L. & Sanders, J. W. (2004): "On the Morality of Artificial Agents", in: *Minds and Machines, 14.3,* pp. 349–379.

Galvan, J. M. (2003): "On Technoethics", *IEEE-RAS Magazine,* 10 (2003/4), pp. 58–63.

Kitano, N. (2007): "Animism, Rinri, Modernization; the Base of Japanese Robotics", *ICRA'07 2007 IEEE International Conference on Robotics and Automation,* 10-14 April 2007, Roma, Italy Full Day Workshop on Roboethics Rome, 14 April 2007 (The world's leading professional association for the advancement of technology). (http://www.roboethics.org/icra07/contributions.html. August 24 2009).

Kitano, N. (2006): "'Rinri': An Incitement towards the Existence of Robots in Japanese Society", *IRIE 2006,* vol.6 (Ethics in robotics), pp. 78–83.

Liddle, B. & Nettel, D. (2006): "HIGHER-ORDER THEORY OF MIND AND SOCIAL COMPETENCE IN SCHOOL-AGE CHILDREN", *Journal of Cultural and Evolutionary Psychology,* 4 (2006) 3.4, pp. 231–246.

Lin, P.; Bekey, G. & Abney, K. (2008): *Autonomous Military Robotics: Risk, Ethics, and Design*, Prepared on: December 20, 2008.

Nakada, M. (2006): "Privacy and *Seken* in Japanese information society: Privacy within *Seken* as old and indigenous world of meaning in Japan", in: Sudweeks, F.; Hrachovec, H. & Ess, C.

(eds.): *Cultural Attitudes towards Technology and Communication 2006*, Murdoch University, Perth, pp. 564–579.

Nakada, M. (1982): "Saigai to Nihonjin (Japanese and natural disasters)", *Nenpousyakaisinnrigaku* (Annual review of social psychology), 23, pp. 171–186.

Nakada, M. & Tamura, T. (2005): "Japanese conceptions of privacy: An intercultural Perspective", *Ethics and Information Technology*, 7(1), pp. 27–36.

Narumi, M. & Imai, M. (2003): "Human-Robot Interaction with Directed Dialogue)", *IPSJ SIG Notes*, ICS 2003 (100), pp. 67–74.

Sanfeliu, A.; Hagita, N. & Saffiotti, A. (2008): "Network robot systems", *Robotics and Autonomous Systems*, 56 (2008), pp. 793–797.

Scassellati, B. (2002): "Theory of Mind for a Humanoid Robot", *Autonomous Robots*, 12, 2002, pp. 15–24. (http://groups.csail.mit.edu/lbr/hrg/2000/Humanoids2000-tom.pdf).

Sullins, J. P. (2007): "When Is a Robot a Moral Agent?", *ICRA 07 – Workshop on Roboethics* (subimitted paper to this workshop).

Veruggio, G. & Operto, F. (2006): "Roboethics: a Bottom-up Interdisciplinary Discourse in the Field of Applied Ethics in Robotics", *IRIE 2006*, vol. 6 (Ethics in robotics), pp. 2–8.

Can Robots Plan, and What Does the Answer to this Question Mean?

Armin Grunwald

Abstract: In this chapter, the question, whether a competence for planning can and should be attributed to robots - as is a common language usage in robotics - shall be dealt with. The question under consideration is, in which manner, with which right, and to which ends one could say that autonomous robots "plan", which understanding of acting and planning it is based on, and which conceptual implications are connected with it. Profound conceptual questions regarding the distinction between human beings and robots are concealed behind these seemingly simple questions. The hypothesis which is to be developed is that - based on the connotations of "acting" and "planning" as attributive terms - it is an empirically substantial question, whether, according to a once agreed-upon concept of acting, robots can be said to act or plan.
Keywords: Planning Theory, Theory of Action, Autonomy

1. The Formulation of the Question and a Hypothesis

Window-cleaning robots replace window-cleaning crews, search engines make inquiries in the Internet, autopilots fly planes, service robots work in chemical or nuclear power plants, automated monitoring systems replace or supplement human guards. Inasmuch as one usually says of the human beings who have done these chores until now – and it can doubtlessly be said of them – that they, in cleaning windows or in servicing technical systems, *act*, it is worth reflecting on, whether, under which conditions, and with which implications this concept of acting can be can be ascribed to robots. This takes place, in particular, against the background of a debate in the sociology of technology, in which, above all, in the scope of the actor-network theory (Latour 1995), an action basis for technology is assumed (Schulz-Schäffer & Rammert 2002) which extends even to a complete action-theoretical symmetry between human beings and technical artefacts.

In this contribution, the question, whether a competence for planning can and should be attributed to robots – as is a common language usage in robotics – serves as an illustrative example (cf. Part 2). In its focus stand "autonomous" robots, which, for example, have to find their way through unknown surroundings, and are not operated by remote control. Because planning is a specific type of action (s. Part 3 on this point,), robots must – if they can *plan* – resp., if this ability is attributed to them – also be able to *act* (resp., consistency arguments would require attributing competence for acting to them as well). The question is, in which manner, with which right, and to which ends one could say that these robots "plan", which understanding of acting and planning it is based on, and which conceptual implications are connected with it. Profound conceptual questions regarding the distinction between human beings and robots are concealed behind these seemingly simple questions (Grunwald 2002).

The hypothesis which is to be developed and substantiated below is that – based on the connotations of "acting" and "planning" as attributive terms – it is an empirically substantial question, whether, according to a once agreed-upon concept of acting, robots can be said to act or plan. Technological progress increases the robots' "autonomy"; for that reason, there can be – due to this progress – different answers to this question at different times. In this manner, a field in which a shift of limits between human beings and technology can be observed, diagnosed, and interpreted presents itself. The term of "attribution", however, has to be stated more precisely. The abovementioned hypothesis is only then tenable, if the attribution is made from the perspective of an external observation, and when questions of a "personhood" are avoided.

2. The Postulation of Planning Robots – a Phenomenology

In speaking about robots, in particular, about autonomous systems, terms from the field of planning are often used. The roboticists' fundamental assumption is:

> "that intelligent performance is possible for computers which are capable of algorithmic pattern processing. Sensors provide input data which are depicted in patterns, and, according to certain rules, are transformed into one another and processed, the result is then used to direct motor "output", in other words, to move actuators" (Schlachetzki 1993: 72).

> "Solving problems, searching, and planning are the means with which instructions ... can be obtained ... out of a large knowledge base. ... it is a matter of defining the initial state, the goal, and any known intermediate states. ... In general, a number of paths lead to the goal. ... In planning, a means-end table is first drawn up in order to recognize central decisions" (Decker 1997: 12).

The classical application is that a robot has to find its way through surroundings unknown to it, for instance, overcoming obstacles in moving forward. Another application concerns soccer-playing robots, for which there is the additional challenge of coordinating the action of several "players". On the part of the constructors, development towards greater robot "autonomy" is expected, which will further increase the demands on planning which the robots will have to do in order to perform the tasks intended for them, as well as to meet the demands on these robots' abilities to plan "autonomously", and to be able to deal with unforeseen, unprogrammed situations:

> "In the future, robots will be far more than special machine tools. They will work together autonomously with human beings to do complex tasks and to do defined work without exact specification of the course of action in unknown or changing surroundings" (Steusloff 2001: 7).

Aspects of planning play a central role in the theory of artificial intelligence and of the realization of "autonomous" artefacts (e. g., Pollock 1995). The construction of an autonomous technical problem-solver on the basis of information processing and sensor-sustained recognition of the situation is one of the central motivations of research on AI (artificial intelligence) and AL (artificial life). A robot as an autonomous system given the task of finding its way through an unknown environment and of carrying out a set task – for example, a transport operation within a building – is one of the most important applications and tests of this performance.

This poses the question of the understanding of planning which applies here, and of the relationship of this understanding of planning to human planning. In conceptual respect, it is a question of whether merely anthropomorphic manners of speech are metaphorically ascribed to technical artefacts here, or whether, as it is said, above all, in some considerations in the sociology of technology, that it will come, through new developments, to an "action capacity" of technical artefacts which can extend even to an assumption of a symmetry between humans and robots. If competence for planning is ascribed to robots in a more than metaphoric manner, they are incorporated into a "community of planners" – a step towards a socialisation of technology (Joerges 2001). The import of such an attribution becomes apparent in comparison with philosophical anthropology, in which the ability to plan and the disposal over the linguistic means for the visualisation of possible futures necessary for planning, was seen as an element of humanity's special position (Kamlah 1973). On the other hand, questions of responsibility are touched upon – who would be responsible for a robot's "plannings" and for their consequences?

3. Planning as Action

Planning is a specific form of action (e. g., Habermas 1968, Grunwald 2000). For that reason, the introduction of the concept of planning begins with a definition of action. On the level of the phenomena, we observe changes, which we categorize partly as a sequence of events, processes, behavior, or actions. The manner of categorisation depends on the criteria and definitions used. The definition of action – in contrast to behavior – given below is used in the following (cf. Habermas 1996, Janich 2001):

- Actions can be attributed to actors (as the cause of the action); this requires assuming the perspective of an observer with regard to one's own actions or those of others;
- Actions can be carried out or omitted, not in the sense of any freedom whatever, but on the basis of reasons; whether this is the case, requires making an interpretation on the part of outsiders, whereby the actor can definitely be asked;
- Actions can succeed or fail, i. e., there are criteria and conditions for success; commonly, these are demonstrated on goal attainment and realisation of the purpose, and can also indicate a partial goal attainment.

The classification of something *as action* is, consequently, an interpretation in a communicatively-structured space. The interpretation is made in reconstruction by external observers, or by the actor him-/herself. In the latter case, it is necessary that the actor dissociates himself from himself, in order to be able to make the interpretation. "Acting" is no ontological predicate, but – in order to be able to speak of acting – an interpretation of the corresponding situation and an argumentatively explicable *attribution* are necessary. If a truck drives by, we don't say that the truck is acting, but that the truck driver is acting, whereby a causal relation between the driver's action (putting on the brakes) and the perceivable effects of these actions (the truck stops) is assumed. Slipping on ice or a coughing bout would not fulfill the criteria for acting, as a rule, but are "occurrences" (Kamlah 1973). A coughing bout can neither succeed nor fail. This is only seemingly trivial, because there are situations in which coughing can be chalked up *as acting*: removing crumbs from the respiratory tract, wanting to attract attention in the concert hall, or warning a business partner that he is in danger of making a mistake in negotiations.

This shows clearly that the classification of a phenomenon *as acting* is done through *attribution*, which is based on an interpretation of each specific situation and of its context (Schwemmer 1987: 83). Two coughs may be phenomenologically identical, can, however, possibly through the interpretation, in the one case be categorized as *behavior* and occurrence, and in the other as *acting*. Inasmuch as interpretations can be controversial, the attribution of the concept of acting can also be challenged in individual cases.

Attributions of concepts are themselves actions, and can be judged according to action-theoretical categories. Ends, e.g., of differentiation, of classification, or of ordering, are pursued with definitions as well. To be able to argue for or against certain attributions, it is necessary to name the objectives they are supposed to realise. Only relative to these purposes can one judge whether the attributions were appropriately chosen or not. At this point, it would also have to be explained, which aims are pursued with the definition of the action concept discussed above.

In the first place, the practical purpose of learning is to be named. It can be derived from the criteria given above, that acting is seen as *capable of being improved*, mere behavior, however, is not. Means-end knowledge, and knowledge for correcting faults can only be referred to action, not to behavior. Acting is subject to conditions for success, out of which a measure of success can be derived, and makes learning possible, because – to express it technically – the target-performance difference of a failure or of a partial success can be used to inquire about the causes for this difference, and to think about and to introduce possible improvements.

In the second place, the question of responsibility has to be asked. With reference to actions, one can speak of reasons, consequences, and responsibilities. Inasmuch as humans describe themselves as being – at least repeatedly – capable of acting, they set themselves in a sociocultural context and define themselves as social beings who can, in preparation, discuss actions, who can choose among alternative options for action, can carry out the actions, and can beforehand as well as afterwards talk about consequences and responsibility (e. g., Grunwald 1999). The concept of action in differentiation from mere behavior is part of the (modern) human being's self-constitution – and it isn't compelling. Cultures are conceivable which don't make this distinction, but which subsume all of this under behavior, and, for example, know no such concepts as responsibility, guilt, justice, and injustice. The distinction between action and mere behavior serves to clarify, to which objects a discussion of responsibility can be extended. It therefore has to be asked in the following (Part 4), whether and to which extent an attribution of a capacity for action or a planning competence to robots would collide with these purposes, or would realise them.

Planning is a matter of active occupation with *future* action for the purpose of consideration and preparation. Planning is an anticipating reflection on purposes/goals or of action schemata without directly positing or actualising them: a drafting of future options for action in the sense of a "test action" (Schütz 1981, Stachowiak 1970). The purpose of planning is the previous drafting, reflection, and judgement of the possible options for acting to the end of preparing the action and its optimisation. Planning leads to purpose realisation only in an indirect manner. Only the plan *implementation* is supposed to realise the purpose, not the plan design itself. Plans result out of planning, while only the *translation of a plan into action* is supposed to reach the plan's goals. Planning is a (hypothetical and experimental) action in the space of possibility and presents itself:

> "as a dramatic testing of different, competing possible directions of acting in the imagination (...) Experimentally, various elements of habits and drives are combined with one another, in order to find out how the resulting action would look in case it was initiated." (Dewey 1922: 190)

Schütz (1971, 1981) analyses, footing on Dewey, the internalisation and course of singular actions from the perspective of a phenomenologist, and likewise emphasises the difference between the draft and the "real" action:

> "(...) devising action takes place in principle independently of all real action. Any drafting of action is much rather an imagination of action, i. e., an imagination of spontaneous activity, but is not the spontaneous activity itself." (Schütz 1981: 77)

Planning is devising and preparing goal systems or action structures which prima facie are neither known nor evident: If one could already do something, i. e., had an action routine at hand, one wouldn't need to plan. Planning is an "intellectual anticipation of future action" and a "method for the attainment of adequate action anticipations" (Stachowiak 1970). Planning doesn't mean habitual action, routinized and customary action, but always concerns only situations with a need for design, preparation, construction, composition, or decision.

An essential attribute of the concept of planning is its *second order purposive rational character* (Grunwald 2000). First, each of the individual action steps of a plan have to be purposive rational in that they should lead to the realisation of certain sub-goals. Second, these elements also have to be arranged in a purposive rational manner. The composition of the, in the particular case, purposive rational elements has to be done so, that the objective as a whole can be attained: Planning consists in the purposive rational composition of purposive rational elements (Habermas 1968). Counterexamples, in which individual, definitely purposive rational elements are combined in a nonsensical manner, can easily be constructed. Alone the often necessary chronological or pragmatic sequence of individual steps leads to restrictions of the possibilities for combining the parts.

This second order purposive rationality implies that planning takes place within the space of reasons and knowledge, and has to be done discursively (Grunwald 2000: Chap. 4). A planning discourse consists of (1) a discourse on the determination of the purposes and goals (*goal planning*), (2) the elaboration of alternative options for scenarios and means, *possible plans* therefore, and (3) the decision for choosing among the alternative options (*decision discourse*).

4. Can Robots Act?

The first part of the question is, whether, through the conceptual definition of acting (s. above), it isn't already analytically predetermined that non-human agents can't come into question as actors. This is not the case, because the model of determining action by means of reconstructively comprehensible criteria leaves open, who who (or what) comes into question as an actor. It is an empirical question, whether the criteria necessary for "acting" can be fulfilled only by human beings, by certain human beings in certain situations, by rational beings in the sense of Kant, by certain animals (e. g., primates) – or even by robots. The definition of acting determined by criteria is open in both directions: not all human beings have to be capable of acting, and actors don't necessarily have to be human.

A look at very young children, at dement individuals, at coma patients, at certain types of disabilities, and at people with compulsive mental disorders shows that not all human beings can act. In the last-named case, for example, the criterion that an action can also be omitted is not fulfilled. Which people are capable of acting in which situations is an – at least: also – empirical question with possibly legally relevant consequences.

Conversely, beings, which don't belong to the species *Homo sapiens*, but can nonetheless act, are at least conceivable. Logically, this can't be ruled out, but is an empirical question of the fulfillment of criteria on the basis of interpretations and reconstructions. Here, it can admittedly come to considerable problems of judgement, when the behavior of a chimpanzee, for example, is supposed to be classified as action. The necessary interpretations can come under criticism as mere anthropomorphising misrepresentations, because humans are not accustomed to maintaining a discursive community with primates. The latter also applies for the relationship between human beings and robots. A difference to the primates, however, consists in the fact that the robot, as a construct of human beings, should be better known in its functioning than primates, and that interpretations, for that reason, could be facilitated.

The definition of acting doesn't limit the competence for action to human beings, and is not speciesistic. If it is claimed that only humans can act, this would have to be the result of empirical analysis with regard to meeting the criteria for acting. This assertion would then be fallible, in particular, because technological progress produces new "beings", which, in principle, could falsify this assertion.

Speciesism with respect to the concept of acting would also be unproductive. If, per defintionem, only human beings could act, then humanity's special status would be conceptually established, with the seeming advantage of eternal consistency, but with the considerable disadvantage of fruitlessness. In the course of technological progress, nothing could be learned about possible shifts of the limits in the relationship between humanity and technical artefacts. Only when the question of acting is an empirical question can historical changes be reconstructed, can one speak of developments and shifts. The practical purpose of making learning processes possible through the definition of the concept of acting, lets itself here be applied to the manner of handling the action concept itself: The purpose is realised only by a definition which permits perceiving and reflecting on changes.

Now, the next step would be to ask how the "robot planning" described in Part 2 appears in the light of the definitions of "acting" given. In short, three criteria bring the following estimation:

- *Action Causation*: There is presumably no question that "autonomous" robots can cause something, in the sense that the effects of their "actions" can be causally ascribed to them. This is, for the capability to act, admittedly merely a necessary, but not sufficient condition, and also applies to mere behavior.
- *Success/Failure*: Inasmuch as such robots have a task (e. g., defusing mines, running errands, or mowing the lawn), success, failure, or partial success can easily be determined from the perspective of an observer.
- *Omission*: A specifically robotic "action", for example, avoiding an obstacle, seen from the perspective of an external observer, could have been omitted, analogous to the act of a human agent – namely, if the arguments which were the decisive factor for the action had been different, for instance, due to a different diagnosis of the situation. Inasmuch as human freedom is not to be understood in the sense of a randomizer, but means the freedom to decide on the basis of good reasons, one would have to concede that a robot which chooses the one option out of spectrum of action schemata which is suited to the diagnosis of the situation and to the tasks it has to carry out, would have desisted from this action and have chosen another, if the reasons had been different.

These considerations don't result in any conceptual grounds for denying "autonomous" robots the capability for acting.

Now, a "consequentialistic" argument, which therefore operates with the consequences of an attribution of action competence, is repeatedly brought forward: the argument of the ascription of responsibility connected with it. First of all, we have to note that consequentialistic arguments don't logically apply here. In case the conclusion is that, according to the culturalistic concept of acting, competence for acting would have to be attributed to robots, then precisely this would be the case, and an argument, which operates with the undesirability of this attribution's consequences would be senseless. Nonetheless, this argument is to be refuted below, because there is something to be learned by it.

With regard to the second purpose of introducing the action concept – using it for clarifying responsibilities (s. above) –, it is occasionally said that the attribution of action competence is, of necessity, connected with the attribution of responsibility, and that this would be contraintuitive. A somewhat closer look at the concept of responsibility shows the short circuit in this argumentation. The concept of responsibility can, first of all, be constructed as a *threefold* concept: *someone* (a responsible subject) is responsible for *something* (result of action as the object of responsibility) to an *instance* (a person, an institution, etc.). The threefold reconstruction is the basis for the type of responsibility which Lenk describes as *action causal responsibility* (1992: 27f., 82). This responsibility reflects on nothing other than the cause of the results of an action by an actor, and has, of itself, no moral dimension. This discloses itself only in the reconstruction of the concept of responsibility as *fourfold*, namely, when one asks *relative to which system of rules* responsibility *should* be assumed (Grunwald 1999). If one argues that the attribution of action competence to robots implies the attribution of legal or moral responsibility, then this is based on a confusion of the three- with the fourfold concept of responsibility. The mere responsibility for causing an action would be applicable to the robot, inasmuch as acting is conceded to it – but that implies no sort of legal or moral responsibility. It is simply the acknowledgement that the robot is "causally" responsible for certain results of the action; to say it with Lenk: The results of the action (for example, that a human being was injured by a collision) can be interpreted as the result of the robot's action. A moral or legal responsibility, however, doesn't automatically follow out of this fact – but its attribution would be an independent act – finally, the broadening of the three- to a fourfold concept of responsibility. Also in the case of an action causal responsibility on the part of the robot, the attribution of the legal and moral responsibility for the bodily harm could lead to the robot's owner, its operator, or its manufacturer (Christaller et al. 2001).

This is, incidentally, very similar also in the case of human action, although we presumably connect the responsibility for bringing about the action – not exactly logically, however – with moral or legal responsibilities. Out of the fact that someone has caused something through his/her action, it doesn't yet principally follow that he/she will also be made morally or legally responsible for it. Instead, this responsibility could be sought by the superior, by the parents, or in the organisation of institutional structures. We are familiar with situations, in which somebody has done something, but someone else assumes the responsibility for it (for instance, the superior officer in the armed forces, who has issued an order), just as well as situations in which someone is held responsible, who has nothing directly to do with the fact in question (for example, a minister who has to resign, because irregularities in his ministry became known, in which he personally wasn't at all involved). These examples make clear that we have to distinguish between the action causal responsibility and legal/moral responsibility. Even if we concede competence for acting and planning to robots, this doesn't anticipate a decision on responsibilities in a moral or legal sense.

The argumentation for attributing action competence to robots presented above is subject to a premise, which gives reason for further differentiation. Understanding "acting" as an attributive term was, in this argumentation, assumed in the perspective of an external observer, who observed "phenomena" such as the human who, or robot which brings about something, and interprets these phenomena according to the concept of action introduced. If a human being was first observed in this manner, e. g., an orderly or a courier, one could – in view of a functionally equivalent replacement of these humans by a robot – imagine that this robot would be "disguised" as a human being, so that an external observer couldn't orient his attribution on superficialities. His external interpretation would – under the assumption of functional equivalence and disguise – have to arrive at the same result for the human being and for the robot. Since the disguise can't correlate with the attribution of competence for acting, this thought experiment would confirm the train of thought developed above: robots, which replace acting humans, act.

The decisive prerequisite is, however, that the human being or robot observed isn't asked about its activities, tasks, diagnoses, and reasons, but that the interpretation is made solely from the observing and reconstructing external perspective. Other questions, for instance, about the actor's self-awareness, or about his status as a person, are not asked. If this is criticised as a deficit, the place in the concept of action would have to be pointed out, in which such questions became decisive for the attribution of acting competence.

5. What Do Robots Do When they "Plan"?

Usually, two types of planning robots are distinguished, according to the differences normally made between AI- (artificial intelligence) and AL- (artificial life) robots (cf. Kinnebrock 1997: 10ff.):

- Types of robots which, on the basis of a diagnosis of environment data, can choose from a predetermined number of options for action, according to a – likewise predetermined – criterion grid.
- Robots based on neural networks, which can incorporate learning effects and then change the basis for planning and deciding.

In both cases, planning as a draft and preparation for future action recognisably plays a role. The first type is of a rather simple nature, because the number of actions and the criterion grid are predetermined, and not changeable. Planning is, in this case, limited to an assignment of options for action to a diagnosis of the situation. The second type, however, is of particular interest, because the action schemata, which come into question are possibly "generated" originally in the robot, and are not previously programmed as part of a predetermined quantity. Learning through an accumulation of "experience" which is made by carying out "actions" – for example, in moving through an unknown terrain – is the heart of this "planning". In it, the "emergence" of new ways of acting which result out of learning processes in an unforeseeable manner is also conceivable (Kinnebrock 1997: 102ff.). This leads, among other things, to the fact that a control architecture is necessary: The unforeseen behavior of a robot needn't consist solely in the generation of *desirable* innovative problem solutions. *Undesirable* results are also conceivable, namely, that robots can "get out of control". The control architecture is supposed to ensure that the robot's behavior remains within a defined frame, and that otherwise the robot is turned off.

If planning is understood to be an experimenting "test action", the question of the more detailed sequence in a robot's learning process poses itself. The explanation:

> "Learning consists of the reorganisation and reevaluation of the individual links within a neural network ... We have previously spoken of supervised learning, by which, for example, human beings exercise control. If we go further to unsupervised learning, then we replace the monitoring system by a set of well-defined rules for learning. The entire system optimises itself according to these learning rules, while we leave it to itself" (Schlachetzki 1993: 78f.)

leads to the conclusion that this is a case of learning through experience. In reflecting on an autonomous robot of the type AMOS (Knick et al. 1994), it is, first of all, important that it disposes over no prefabricated model of its surroundings, but produces one itself through experience, and continuously improves and adapts it. Illustrated on the example of dealing with obstacles (one might think, e. g., of a delivery messenger or courier robot in an administrative body): Through sensor signals, the robot generates a model of its surroundings while it is moving. As long as these surroundings are static, the model produced (consisting of walls, doors, elevators, etc.) can be used without problems. During operation, the robot constantly checks, by means of sensory technology (e. g., video cameras), whether its model is still up to date. If a door, which is normally open is once, by chance, closed, it comes to a "planning breakdown", just as when an obstacle unexpectedly lies in the way. Plan breakdowns designate deviations of the real situation from the expectations. In such cases, the robot defines the area in which a difference between the model of the surroundings and reality occurs as a "region of interest" (ROI, Knick et al. 1994: 77ff.). Through experimental handling of the unexpected situation, the robot can gather "experience". It can try to bring the obstacle to make way by sounding a horn (the obstacle could be a human being who steps aside after the horn sounds), it could try to push the obstacle aside (maybe it is an empty cardboard box), or the robot could, if nothing else helps, notify its operator. Maneuvers such as parking or making a U-turn in a corridor can be "planned" in this manner (Schlegel/Illmann 1995). One of the most important challenges in this work is classifying the plan breakdowns (Knick et al. 1994: 80ff.), in order to be able later to diagnose the right type and to take the appropriate measures (if a human being is standing in the corridor, to sound a horn instead of trying to push him/her aside).

The underlying planning-theoretical paradigm consists, in essence, of the cybernetic planning model of feedback in a system-environment interaction: The heart of the planning concept in the systems-theoretical approach (e. g., Stachowiak 1970, Chadwick 1978) consists in the *implementation* of the plan and, in the form of cybernetic feedback, includes the *examination of the results* (Stachowiak 1970: 4; Chadwick 1978: 375ff.). Planning consists accordingly in a "system's"dealing with its surroundings: A system takes measures for the change of the system-surroundings desired (experiments). Errors are detected by means of the feedback control mechanism, and are taken into consideration in other measures. Learning consists in repeated runs of this cybernetic loop, with a corresponding accumulation of empirical information. A robot's experimenting with unknown surroundings and the use of the resulting "experience" can, in fact, be interpreted as planning processes within the framework of cybernetic planning theory

The reference to the concept of planning introduced above, the specifics of purposive rationality of the first and second order, as well as the coherence, the "tree structure" of action plans, and the necessity of distinguishing among alternative options (Grunwald 2000: Chap. 3) are no grounds for rejecting the concept of the "planning" robot. The robot makes – through sensors – an interpretation of its present situation and compares it with a (strict, or within a certain range predetermined) goal situation. Out of a knowledge base, it compiles, by means-end relations, scenarios and possible plans of action, and decides on the choice and composition according to a (predetermined) criterion grid. The specifics of planning, especially the purposive-rational draft of actions yet to be carried out, are here recognisably included. Planning-theoretical modelling by the robot through the cybernetic loop is, therefore, possible and adequate.

In the case of this robot planning, these are, of course, in comparison with the complexity of human planning (Grunwald 2000) extraordinarily limited planning processes. This is to be elucidated in two directions: (1) by exposing the cybernetic feedback as a deficient planning model, and (2) by examining the pre-planning agreements.

(1) The cybernetic mechanism consists in learning from experience through more or less well preformed testing and practical trials of action steps in the cybernetic control loop. In the model of an "adaptive continual planning", the robot adapts itself to the conditions of its surroundings. The normativity of planning – namely, to make a plan according to certain determinations of aims and, if applicable, to implement it – is neither taken into account in the cybernetic model, nor is there a mechanism which could reconstruct this normativity there. The mechanism of checking the results of planning and of the comparison of the present situation ("is") with the desired one ("ought") simply acts as a substitute for the normativity of the determination of the planning's goals. The robot isn't compelled to determine objectives beforehand, to reason about means for reaching the goals, and about eventual incidental consequences, but it can – on a configured initial basis – try actions out, and classify the results. What is without doubt sensible in the case of, for instance, AMOS (s. above), fails, however, in planning tasks of other types, such as building a house, or of large-scale technical projects. Instead of adaptation to environmental conditions, it is, in the latter case, a question of defining objectives and of realising them. The specifics of this type of singular planning (to which also such highly complex plans as the Apollo mission to the moon belong) are the normativity involved and and the previous reflection or even modelling and simulation of the entire process. In comparison, cybernetic planning is nothing more than an improved method of trial and error – a method which, as a rule, plays no great role in normal human planning.

(2) A second type of limitation of robots' competence for planning results out of the question on the decisions made before planning. Concrete planning isn't free of premises, but is based on preliminary decisions, through which the space for possibilities,

options, and searching for the solution of the respective planning task is predetermined. The type of task formulation, performance targets for planning strategies, and the initial conditions to be observed are *selective*. They limit the manner in which one can plan, and how possible plans could look. These perliminary decisions are *pre-planning agreements* (Grunwald 2000 Chap. 4.2). Because of limited time or other resources, *all* of the conceivable alternatives for solving a planning task can never be taken into consideration. Much rather, solutions from a certain frame of relevance and spectrum of options are always, explicitly or implicitly, permitted. Elements of such a *relative planning a priori* are:

- narrowing down the object of planning in the form of a determination of the target areas to be taken into consideration, the array of objects, and of the relevant context in modelling and the determination of the system's limits;
- the "state of the art" as the scope of the available planning- and context knowledge as the initial situation for planning;
- decisions regarding the admission or prohibition of goals and means (e. g., for religious reasons), and
- criteria for decisions on the choice of a plan among several possible ones, resp., determination of the decision-theoretical meta-rule for the process of selection.

Pre-planning agreements of this type are contextual restrictions of the principally conceivable diversity, and are elements of contingency reduction. Planning contexts can, in the scope of a "relative" planning's a priori, be distinguished according to whether and to which extent the pre-planning agreements are under the planners' control, for example, were negotiated before beginning to solve a planning task, or whether they were set for the planners "from outside".

A robot's planning is, in fact, as portrayed, *reproducible, resp., describable* in the cybernetic planning model. In it, the objectives stated and the target setting are limited; in part, algorithm-like sequences are determined, in part, the knowledge basis is predefined, limits are set for the robot through the control architecture, and so on. Markedly restrictive *pre-planning agreements* have been met which *can't be revised by the planning robot*. The robot's "behavior" can, in

fact, definitely be designated as "planning"; it is, however, – in spite of all of the possibilities of an experimental classifying and adaptive learning – a very special and reduced type of planning:
"The behavior of autonomous robots is – by using the techniques of information processing available today – marked by their knowledge basis in the form of programs and data, and precognition in which representation whatever. This knowledge basis, its use and enlargement is, even in the case of so-called self-learning systems, predetermined by humans in the realisation of robot systems". (Steusloff 2001, p. 7).
We now have to take into consideration that the simple comparison of a "freely" planning human being and of an, after all, to a great extent controlled, robot planning falls short. Human planning also takes place in strongly restricted possibility spaces (e. g., within restrictive employment relationships). It seems, and this would be the résumé, that there is a *gradual transition* from the planning of a simple robot with restrictive planning agreements to "free" and complex planning processes, which doesn't necessarily make a qualitative jump at the transition from robot to human being.
In this manner, it becomes possible to reconstruct shifts of limits. Inasmuch as technological progress will increase robots' "planning competence", the previous limits will be shifted. The borders between humans and technical atrefacts are becoming blurred in this area as well. Technology is being developed in a manner, which allows technical artefacts to perform increasingly complex activities. This should be well reconstructible in in planning models when the robots' decision-making powers, and therefore their autonomy, grows through technological progress.
Latour's demand (1995), to speak of robots and human beings in the same language, and to acknowledge a complete symmetry between them, turns out to be not very helpful in this connection. For the concept of planning, it is, in fact, possible, as has been shown above, to speak of planning robots as well as of planning humans, from the perspective of an external observer. A complete symmetry between humans and artefacts is *just precisely not* connected with it. Use of the same expressions doesn't mean any acknowledgement of symmetry, but the conditions for the use of the language also have to be reflected upon (e. g., the difference between a threefold and a fourfold concept of responsibility, as shown above). Concluding from the use of the

same action- and planning terminology to a symmetry between humans and robots would be possible only by an extreme disregard of the different planning models, the differing disposal over pre-planning agreements, and of the different treatment of the normative level. More detailed analysis has shown precisely that differentiations have to be made in order to arrive at a "better understanding" of planning robots and human beings. Only painstaking deliberation and consideration of the differences is instructive: One can, in the comparison of planning robots and planning humans, also learn something about planning humans, namely, specifics of human planning and their in part very narrow limits, set by certain terms of reference, e. g., in employment law.

When we, in reflecting on planning robots and humans, actually do the work of translation and imparting and, in doing it, re-"construct" ourselves (Joerges 2001, p. 196, with reference to Latour), resp., the differentiation between humans and technical artefacts, this functions in particular when we make distinctions. The fact that there are apparently models of planning by the use of which we can speak of planning robots, doesn't mean that the planning of humans and of robots is to be put on the same level. When we assign the attribute "planning" to humans and to robots equally, we still don't necessarily connect the same conceptions with both of them. But rather in each individual case, we apply in this attribution a certain understanding of planning and a specific planning-theoretical model. In a certain sense paradox, the use of the same terms for planning robots and human beings intensifies the asymmetry, instead of bringing about symmetry.

If we reconstruct the work of a messenger robot, we will – formally – find the same action-theoretical structures as when we reconstruct the action of a human messenger *as a messenger*. The putatively strongly-limited objective-setting competence on the part of the robot (in which the tasks are programmed) is no counterargument, because, in an environment regulated by employment law, even the human messenger has a very restricted objective-setting competence; in principle, he has to do what his superior demands, in the scope of his job description, to which he has consented. On this level, the activities of the messenger and of the messenger robot are equivalent – which is also logically conclusive – otherwise, the messenger robot couldn't replace the human messenger. And in spite of this, there is a considerable asymmetry between the messenger robot and the messenger. A robot, which is functionally equivalent to the human messenger, i. e., which brings the same messenger performance, plans the errands and the solution of problems occurring in them in a specific sense and under predetermined initial conditions. The human messenger plans in that he fills his role, according to an analogous understanding of planning, and with probably similar criteria. While the messenger robot, however, is committed to its role as a messenger through programming and control architecture, the human messenger can abandon this role. The requirement for the ability to omit acting in order to distinguish action from behavior has to be differentiated. For the robot, it is already fulfilled when it has the choice between some few alternative options for action – but yet, it remains within its role. The human messenger, on the other hand, can understand the "omission" much more radically, and abandon his role. The measure of the ability to omit action proves to be central for the distinction between human and robot, and also to be a parameter for "measuring" future shifts of the limits in this field.

References

Chadwick, G. (1978): *A Systems View of Planning*, Pergamon Press, Oxford.

Christaller, T.; Decker, M.; Gilsbach, J. M.; Hirzinger, G.; Lauterbach, K.; Schweighofer, E.; Schweitzer, E. & Sturma, D. (2001): *Robotik. Perspektiven für menschliches Handeln in der zukünftigen Gesellschaft*, Springer, Berlin.

Churchman, C. W. (1968): *The Systems Approach*, Dell Publishing, New York.

Decker, M. (1997): *Perspektiven der Robotik. Überlegungen zur Ersetzbarkeit des Menschen*, Graue Reihe 8 der Europäischen Akademie, Bad Neuenahr-Ahrweiler.

Dewey, J. (1922): *Human Nature and Conduct*, Modern Library, New York.

Grunwald, A. (1999): "Verantwortungsbegriff und Verantwortungsethik", in: Grunwald, A. (ed.): *Rationale Technikfolgenbeurteilung. Konzeption und methodische Grundlagen*, Springer, Berlin, pp. 175–194.

Grunwald, A. (2000): *Handeln und Planen*, Fink, München.

Grunwald, A. (2002): "Wenn Roboter planen: Implikation und Probleme einer Begriffzuschreibung", in: Rammert, W. & Schulz-Schaeffer, I. (eds.): *Können Maschinen handeln? Soziologische Beiträge zum Verhältnis von Mensch und Technik*, Campus Verlag, Frankfurt, New York, 2002, pp. 141–160.

Habermas, J. (1968): *Technik und Wissenschaft als Ideologie*, Suhrkamp, Frankfurt.

Hartmann, D. (1996): „Kulturalistische Handlungstheorie", in: Hartmann, D. & Janich, P. (eds.): *Methodischer Kulturalismus. Zwischen Naturalismus und Postmoderne*, Suhrkamp, Frankfurt, pp. 70–114.

Janich, P. (2001): *Logisch-pragmatische Prodädeutik*, Velbrück, Weilerswist.

Joerges, B. (2001): "Technik – das Andere der Gesellschaft?", in: Ropohl, G. (ed.): *Interdisziplinäre Technikforschung*, Hanser, München, pp. 165–180.

Kinnebrock, A. (1997): *Künstliches Leben. Anspruch und Wirklichkeit*, Oldenbourg, München.

Knick, M.; Schlegel, C. & Illmann, J. (1994): "AMOS: Selbständige Generierung bedeutsamer Wahrnehmungsklassen durch ein autonomes System", in: Levi, P. & Bräunl, T. (eds.): *Autonome mobile Systeme*, AMD 94, Springer, Berlin, pp. 77–88.

Latour, B. (1995): *Wir sind nie modern gewesen*, Akademie-Verlag, Berlin.

Lenk, H. (1992): *Zwischen Wissenschaft und Ethik*, Suhrkamp, Frankfurt.

Pollock, J. L. (1995): *Cognitive Carpentry*, MIT Press, Bradfort.

Schlachetzki, A. (1993): "Künstliche Intelligenz und ihre technisch-physikalische Realisierung", in: Verein Deutscher Ingenieure (ed.): *Künstliche Intelligenz. Leitvorstellungen und Verantwortbarkeit*, VDI-Report 17, Düsseldorf, pp. 72–82.

Schlegel, C. & Illmann, J. (1995): "AMOS: Beherrschung vielfältiger Anforderungen durch dynamische Kombination und Konfiguration einfacher Mechanismen", in: Dillmann, R. Rembold, U. & Lüth, T. (eds.): *Autonome mobile Systeme*, AMD 95, Springer, Berlin.

Schütz, A. (1971): "Das Wählen zwischen Handlungsentwürfen", in: Ders.: *Gesammelte Aufsätze, Bd. 1, Das Problem der sozialen Wirklichkeit*, Nijhoff, Den Haag, pp. 77–110.

Schütz, A. (1981): *Der sinnhafte Aufbau der sozialen Welt*, Frankfurt.

Schwemmer, O. (1987): *Handlung und Struktur*, Suhrkamp, Frankfurt.
Stachowiak, H. (1970): "Grundriß einer Planungstheorie", in: *Kommunikation* VI/1, pp. 1–18.
Steusloff, H. (2001): "Roboter, soziale Wesen", in: Kornwachs, K. (ed.): *Tagungsbericht der Gesellschaft für Systemforschung*, Karlsruhe, p. 7.
Strube, G. (1993): "Die Rolle psychologischer Konzepte in der Künstlichen Intelligenz", in: Verein Deutscher Ingenieure (ed.): *Künstliche Intelligenz. Leitvorstellungen und Verantwortbarkeit*, VDI-Report 17, VDI-Verlag, Düsseldorf, pp. 83–93.

Between Innovative Forms of Technology and Human Autonomy: Possibilities and Limitations of the Technical Substitution of Human Work

Peter Janich

Abstract: Human cognition is intrisicly connected with intentions of human beeings. Cognitive machines, above alle those of processing linquistic forms of data, have to suspend the intentions incorporated in linquistic utterances made by a human being towards another one in order to technically substitute the performance of communicating persons in an equivalent way („leistungsgleiche Substitution"). Some types of language machines are discussed as to cleraify the constraints and limits for suspending intentionality and shifting it to the construction of the machine and its program. The background of this methodical approach is the problem of performing semantic issues by means of syntactic machines.

Keywords: cognitive machines; human autonomy; language processing; artificial intelligence, methodical philosophy

0. Introduction

This essay aims to clarify the possibilities and limitations of technical systems that substitute human work by 'performing equivalent work'. The focus will be on *cognitive machines*. One sub-set of cognitive machines is made up of those that process human language. Here it will be appropriate to consider how the *intentions* of speakers and listeners that are crucial to human communication can be dealt with in the context of the equivalent substitution of human cognitive accomplishments by technical systems. To this end a list of 'hermeneutic specifications' will be drawn up to define how 'equivalence of outcome through suspended intentionality', i.e. suspension of the intentions that humans have when speaking and listening, can be achieved through technical means.

The philosophical means to this end will be a clarification of terms within the framework of a methical approach, or to be more precise a theory of action approach[1]. The essay will show what exactly needs to be taken into account in the design of such machines. It will not explicitly pursue the agenda of 'theory of science as critique of science', according to which familiar theories of language processing

---

[1] Concerning the philosophy of language and theory of action underlying this essay, see Janich 2001.

and robotics are subjected to methodical analysis and critique. It will rather constructively explore criteria for extending the boundaries of language-processing machines.

The structure of the discussion will include a *first* section containing analytical descriptions of types of techniques that substitute human work. A *second*, critical section will refute established philosophical dogmas such as naturalism, materialism, causalism, empiricism and others, in order to retain a focus on the human agent as the responsible author of all the technologies to be discussed here. The *third* section will constructively clarify terms, and identify the investments made by theories of action and philosophy of language in distinguishing cognitive and linguistic accomplishments. Referring to that, the *fourth* section will deal with the limits of substitutability. The *fifth* section will once again take a specialised approach to discussing machines that understand and speak, highlighting the growing divergence between the equivalence of outcome and the functional equivalence of what human beings and machines 'do'. The *sixth* section will then go on to discuss 'suspended' intentionality in technical systems. This will allow us to formulate a new type of technology with a new type of avoidance of malfunction or elimination of malfunction as functional criteria, so-called 'hermeneutic machines'. A summary will then conclude by presenting a 'list of hermeneutic specifications'.

1. Classifications (Analytical-Descriptive Section)

'Technology'[2] in the sense of artefacts manufactured by humans for rational ends is not restricted to the substitution of human work. Buildings and furniture, vehicles and roads, and clothing and food are important examples of this. By contrast, machines that 'substitute' human work are those that either make easier or do on our behalf all

---

[2] The common use of the word ‚technology' is sloppy in English as well as in German. It ignores that the ending ‚-logy' litterally refers to a theory or at least to ‚talk about something' and therefore neglects the difference between an object and its liguistic representation. Artefacts, tools and machines are no liquistic but concrete, material objects, of course, which better should be called „technical" than „technological". This matters even more in a philosophical text where the choice of an approriate linguistic representation of non-linguistic items is the epistemological key question. Therefore the meaning of ‚being brought about by a non-lingustic human action' (handicrafting) is called ‚technical' following Aristotle's *physics* where the Dativ *techne* simply means artificial, non-natural.

those jobs that are laborious, burdensome, dangerous, or too complex or costly. The key examples in our life world are household appliances such as dishwashers, vacuum cleaners, lawn mowers, sewing machines and electric tools for the DIY enthusiast. Simple tools such as a shovel can also make manual labour easier or act as a substitute for the human hand, and this technology can also be refined into the more specialised forms of the mechanical digger.

It is assumed that the reader is aware that as these machines become technologically more sophisticated they also become both more specialised and more effective. This is achieved by deviating in functional terms from the models that nature offers. Efficient machines 'do things differently' from people. This is already demonstrated by the basic elements of mechanical systems that incorporate wheels, gears, thermal engines etc.

One particular type of machine, and one that is important for the debate on the possibilities and limitations of robotics, is machinery designed to facilitate or substitute human cognitive labour. Historically speaking, the oldest examples are calculating aids. Established areas today include office machines for administration and organisation, control, regulation and monitoring systems, all forms of calculation, and finally language-based systems such as dictation systems and translation machines, as well as machines providing information services or automated access to expert systems. Finally, one broad field is the use of cognitive machines in the natural sciences and engineering disciplines to perform traditional tasks of measurement, observation and experimentation.

For the purposes of the present essay it is important to note that (1) not all machines which substitute human work by performing equivalent work are cognitive machines, and that (2) not all cognitive machines that make perceptual and cognitive work easier or substitute it are 'language-based systems in the strict sense'. The qualification 'in the strict sense' is designed to distinguish everyday language and specialised languages that are spoken, from artificial and formal languages that are specially suited to processing by machines, chief among which is mathematics. Due to the simple definition of the criteria for what they do, calculators form a special case among the cognitive machines.

The present essay will not cover the use of cognitive machines in the context of scientific research.

2. Refuting Philosophical Dogmas (Critical Section)

Attempts to naturalise the human being (including her brain, mind and soul) and ultimately all her complex cognitive (and emotional) accomplishments by looking only at the causal relationships within material systems that can be controlled empirically in experimental settings are the wrong way to proceed. The cognitive accomplishments of human beings are not the causal outcome of brains or organisms, just as they cannot be the causal outcome of machines, computers or computer chips. This is because *cognitive accomplishments* are *by definition* exclusively those that can and must be judged by criteria of truth and falsehood. It is categorical nonsense, though, to assume that causal relationships and effects are either 'true or 'false'.[3] Here we need to distinguish between on the one hand what a human being accomplishes or the *outcome* (formulated in language) of what a machine does that is substituting her in some equivalent way, and on the other hand the *functional* dimension of the procedure or method applied as a means to achieve this outcome as an end.

It is helpful here to recall the distinction between 'models for something' and 'models of something[4]. A 'model of something' involves a relationship between the model and the modelled that is representational, and in which the two – the model and the modelled – are *described in the same language*, i.e. are of the 'same categorical order' (as for instance with a model railway and the railway itself). 'Models for something', on the other hand, do not model who or what it is that is doing something, but what it is that they do. Here is a simple example. A pocket calculator is required to produce correct results for calculations, but it is not required to go about it in the same way as a human being who is skilled at mental arithmetic, or who works it out using pencil and paper. In other words the pocket calculator is a 'model for' the human calculator, whom it replaces by producing *correct results*. The mechanical calculator is not expected to perform an operation that is the functional equivalent of what the human calculator does. Put more simply, the machine is not required to 'go about things' in the same way as the human being. Generally speaking people are aware of this fact. What fewer people are aware of, though, is that there cannot be either any scientific/technological

---

[3] See Janich 2000.
[4] See for instance Gutmann 1995 and Janich & Weingarten 1999: 88f.

functional description of the calculator, or any causal explanation, indicating that the results of the calculation *are valid (i.e. true)*.

The argument[5] for this that has been put forward for years in methodical philosophy is simple, and takes account of the real and tacit restriction that only functionally undisturbed calculators (or to put it naively, only calculators that are working properly) deliver correct results, while those at are disturbed do not. Now if the validity of the result of the calculation were a causal outcome of the machine's action, then – conversely – false results produced by a defective machine would imply a falsification of the respective causal relationship. De facto (and reasonably), however, no one questions the 'laws of nature' on which the design and functioning of calculators are based. It is rather the case that the *falsehood of a result* is *explained* by identifying a fault and *confirmed* by carrying out a corresponding repair. Malfunctions are deviations from the intended purpose of a machine. In other words, false computational results delivered by a defective machine do not lead to the falsification of an empirical proposition. Instead, they demonstrate that the machine is *failing to fulfil the purpose* for which it was designed, built and operated.

Methodical philosophy succeeded in showing that it is merely the dogmatic restriction of logical-empiricist approaches that considers only logical-definitoric or empirical-causal relationships between the functional medium and the outcome, between the calculating machine and the result of the calculator, as being susceptible to scientific study.[6] However, since machines are artefacts, i.e. products made by human beings in order to achieve certain ends, which therefore possess significant non-natural qualities, we also need to take into account in our scientific analysis (which includes the natural sciences) the means-ends rationality of human action, as well as the causal structure of natural and technological objects and events.

Wherever machines that substitute human cognitive accomplishments are invented, designed, built and used, then according to the *principle of methodical order*[7] a first step is always to define these

---

[5] First put forward in: Janich 1993: 39–53.
[6] The history, systematic exposition and critique of these misunderstandings is developed in: Janich 2006.
[7] The principle of methodical order is the key criterion on which the distinction between methodical and analytical philosophy of science is based. For an exaplanation and justification of this position see for instance Janich 2001.

accomplishments, including the criteria for their validity, as ends to be incorporated into the design. The definitions must be comprehensive enough to enable the machines to determine whether the ends have been achieved or not. Let us once again consider the simple example of the pocket calculator. It must possess not only developed powers of calculation, but also depends on a meta-linguistic competence enabling the constructor or user to *judge*, by applying criteria of validity, the validity or invalidity of schematically generated results of calculations. Only then can the idea of substituting this human accomplishment by a machine make sense.

In other words, how much has been achieved in the invention of calculating aids from the abacus and the (logarithmic) slide rule, through mechanical and electronic calculators, and on to modern mainframe computers, is always linked to the question of how an arithmetic whose validity criteria have already been defined in relation to the externally defined overarching end of producing valid results, is realised technologically through the selection of appropriate means. The inventor and designer of calculators possesses, in the precise sense of the term, autonomy in the definition of this end or purpose. Things are, however, more complicated than this.

3. Intentionality and Cognition

(A preliminary remark from the perspective of the philosophy of language: As we have known since the *linguistic turn* in the theory of science, juggling with nouns such as 'intentionality' or 'cognition' is hazardous and may cloud matters, due to the usually false suggestion of reification. We will therefore now use only the adjectives 'intentional' and 'cognitive'.)

In what way are the human cognitive accomplishments that are to be substituted by machines dependent on the intentional disposition of the human beings who perform them? How can we define the term 'intentional' as simply and as suitably as possible for the purposes of answering this question?

It is imperative not only in everyday speech, but also in specific terminological situations, e.g. in legal and moral contexts, to draw a distinction between human 'action' and a natural event (which may happen to a human being). The latter is usually termed 'behaviour', in the sense in which stones and plants can 'behave', but not in the sense of behaviour as action, as in 'How does a shoplifter behave when caught red-handed?' (A similar terminological distinction exists between 'behaviour' as used in the non-intentional sense, and 'conduct', which is always intentional.)

Non-intentional behaviour that occurs as a natural event can be studied and explained causally, whereas actions are learned in human communities of action and speech, not infrequently as a modification or improvement of natural, innate behaviours. (The examples of walking, swimming, cycling, eating with a knife and fork, speaking, writing, drawing, reading etc. spring to mind here.) In these situations the individual learns to distinguish between actions that others will credit or blame her for, and processes such as growing, digesting, falling asleep, dreaming, tripping over, feeling frightened etc. that are excluded from sanctions.

In everyday speech, we describe actions as 'intentional' when they are performed 'deliberately' by an individual, as opposed to being performed unintentionally or accidentally, or when something occurs naturally or by itself. In legal contexts a distinction is drawn between 'premeditated' and 'negligent', and once again between these terms and 'uninvolved' and 'innocent'. It is a reflection of our cultural achievement that we assume that people (at least those in our own environment) possess sufficient social skills to distinguish between action that is bound by responsibility, and non-intentional behaviour that is not. These definitions of intentional, deliberate, premeditated etc. are problematic, however, when applied to information provided by actors themselves, for instance where it is claimed that an actor could have refrained from an action or performed a different one. The validity of self-disclosures of this kind cannot be verified, because these acts of disclosure are merely honest or dishonest statements, and not true or false propositions.

The concept of action on which we are basing our arguments here is, in other words, not an authenticistic one, i.e. is not based on authentic self-disclosure; it is an *ascriptive* one, which is to say it is based on other people's *ascriptions*[8] to an action performed by an agent. The distinctions between success or failure in the execution of action (whether or not it is properly performed or 'comes off' as intended), and an action's instrumental success or failure (i.e. whether or not its

---

[8] For a detailed exposition of the distinction between ascription and description, see Janich 2010.

end is achieved) also serve to differentiate the incumbent responsibilities: whether or not an agent's action comes off (e.g. as intended), and whether or not she achieves her end (which we usually describe as a learning experience) is something that happens to the agent. But the agent is *responsible* for defining the end and (rationally) selecting the instrumental means to achieve it. In case of conflict, she will be called to account for why she did what, i.e. what she did and to what end she did it, and perhaps also how she did it. If we consider that ends consist in bringing about, avoiding or maintaining states of affairs, which in turn are to be represented by propositions, then we can apply the following general definition: actions are (always, by definition) 'intentional'; moreover, anything that contains an element of action is 'intentional', and vice versa.

One of the most important forms of action is *language*. We also hold each other responsible for this kind of action. Linguistic actions form the most important subset of the 'practical', so called since Greek antiquity (as opposed to the 'kinetic' and 'poietic', i.e. non-linguistic actions of moving and manufacturing). Praxis is everything by which a person affects the needs, interests, desires, dislikes etc. of another person. This can also be achieved by non-linguistic means, as demonstrated for instance by the actions of giving gifts, stealing, injuring, healing, rewarding and punishing etc. When it comes to the substitutability of human linguistic actions by machines, though, it is the communicative linguistic praxis of human beings that makes the crucial difference.

Ascription becomes more complicated with actions that are *only executed successfully with the involvement* of another person, or that *only successfully achieve their end with participation* by another person. Examples of such (non-linguistic) 'acts involving others' or 'interactive actions' include racing or assembling roof timbering. Of the linguistic actions, the holding of a conversation is one example that can only be construed as an interactive action involving another or others. Conversation is distinct from dialogue. ('Dialogue' literally means talking something through or to an end, i.e. so as to reach a decision; a 'successful' dialogue is one in which the dialogue partner agrees with the proposition(s) put forward by the proponent.) But which of the parties to a conversation or a dialogue is responsible for what?

Actions that are all without exception described as 'intentional' are by definition termed 'cognitive', if they aim to achieve results that can expressed through language, and as such can be distinguished as true or false (or the like, such as undecidable, meaningless, definitively refuted etc.).

Using these theoretical distinctions concerning action, we can now return to the issue of the substitutability of humans by machines on a like-for-like basis. We note that the (idealised) human being, when acting as a designer, builder or user of such machines, is also autonomous in the definition of ends, rational in the choice of means and responsible for outcomes. This means that machines as artificially (technically) manufactured objects (or 'artefacts') are dependent on human intentions with regard to both the definition of their ends and their own nature as means. They are not objects that occur naturally. Their artefactual nature always displays non-natural aspects that are determined by the definition of ends and selection of means.

4. Limits of Substitutability

In an earlier work that formed part of a project on manned space travel I showed what constitutes the *fundamental limits* of the substitutability of a human being by machines.[9] This thinking emerged from a report by an astronaut who had to conduct experiments during his mission. A small splint attached to an expensive and technologically complex experimental apparatus got broken, as a result of which the experiment no longer 'worked'. The astronaut then used a nail file to make a substitute splint from a prong on his fork, enabling him to complete the experiment successfully. Would it be possible to design an experimental robot that could substitute this creative human action 'in an equivalent, like-for like way'?

Evidently it would not, because in order to be able to design a machine of this kind it would be necessary to foresee all the unforeseeable incidents, so that the right responses could be built into the design. However, this would mean accepting a conflict of objectives that might even constitute a logical contradiction: attempting to foresee rational goal-driven responses to unforeseeable incidents is attempting the impossible.

---

[9] Janich 1993: 22–26.

We also need to remember this when we consider cognitive machines, and a fortiori when we consider communicating machines that process language. This is because *all communication is characterised by the unpredictability of the response of the communication partner.* If a speaker in whatever situation could regularly predict the response of the listener with mechanistic determinacy, then communication would have failed to achieve its specific end. Linguistic communication is always in some way a process of negotiation between mutually independent individuals who are parties to the communication, for whom we must presuppose a sufficiently symmetrical degree of autonomy. This brings us to the issue of language machines.

5. Language Machines

When we discuss 'language machines' or 'machines that use language', with the connotations of comprehension and intelligent language use that this entails, we should avoid certain widespread inaccuracies and misunderstandings that are common in everyday language (as well as the forms of these inaccuracies and misunderstandings that occur in the academic disciplines). Let us therefore now enumerate four types of language machine based on familiar examples:

Type 1: The highly efficient machines for the *transport and storage of information* (based on the 'mathematical theory of communication' or 'mathematical theory of information' of C. Shannon und W. Weaver)[10]. These include the tape recorder, the CD player, the telephone etc. These machines *do not offer any analogy* to 'speaking' or 'listening' and 'understanding'. They merely *process sound events* by applying criteria of structural preservation. At no point in the system comprised of the five components (information source, transmitter, channel, receiver, information sink) is it necessary to make use of the fact that the sound events in question are those produced and received by human beings in the process of communicative speech. Music or indeed any other form of noise is also mastered by this technology without distinction.

Type 2: The speaking clock telephone service differs from the type 1 machines in that the listener requires the speaking clock to *tell her the correct time*, which means that the service provider also ensures that the speaking clock does so. This example has been selected to

---

[10] See reference in footnote 5.

Between Innovative Forms of Technology and Human Autonomy 221

demonstrate that it is only the parameter of time that determines the accuracy of the sound event coming from the source which the human listener 'understands'.

Other telephone information services differ only slightly, but are fundamentally the same. These include information on current local cinema schedules, the most recent lottery numbers, the latest traffic jam reports from motoring organisations, or today's weather report. The precise criteria, by which any of these services is deemed to be supplying the correct information, are known to any speaker of everyday language. The machine merely needs to reproduce a sound event at a certain time and in a certain place (or within a certain local network).

Type 3: *Interactive information services* in which the machine generates a spoken answer to a question posed by a caller, for instance concerning train timetables. This differs fundamentally from type 2 in that an analogy to 'understanding' must be achieved by technological means. The same principle applies to machines that need to sort items of mail by 'reading' the postcode.

However, we need to bear in mind that this technological analogy to 'understanding' only works on the basis of a comparison of patterns (we will not discuss more simple cases in which the customer 'answers' the 'questions' generated by the machine by pressing keys on the telephone). This is a problematic area, which John Searle discussed using his example of the 'Chinese room'. Ultimately the question is whether a 'semantic' understanding of language can ever be completely substituted by a syntactic understanding – as in the case of a person who has no written or spoken command of the Chinese language, but who can nevertheless by comparing Chinese characters judge whether they are 'the same' or 'not the same', and thus 'communicate' with the being in the Chinese room. The answer to this question is no.

If the timetable information provided automatically is to the caller's satisfaction, by virtue of the fact that it is understood (like a human speaker would be understood) by the caller, and then proved to be correct as the person undertaking the journey catches their train, this is not to say that the machine has 'understood' the caller in the sense in which human speakers and listeners understand each other when they communicate through speech. It is merely similarities between sound patterns that have led the machine to 'respond'. There is nothing remotely like an analogy here to recognition by a human listener of what a human speaker *intends* through her speech. It is rather the case that the end or purpose of the machine, which is to functional well as a provider of information on train timetables, remains external to it, in accordance with the specifications of its designer and operator. Moving closer toward the form of understanding found in communication between human language users, we will now look at another type of language machine:

Type 4: More than a syntactic command of patterns is demanded of *translation machines* that translate written or even spoken texts from one natural language into another one 'appropriately', i.e. in a way that preserves the meaning. They are expected to do more than compare incoming patterns with their own stock of signs and judge whether the two are (sufficiently) similar. This expectation exists regardless of the fact that translation machines are computers, and as such syntactic machines. However, a machine of this kind that can do no more than compare syntactic patterns is like a poor human translator, who translates the text in front of him, but doesn't bother to make sure that the translation is appropriate or that it is understood by the target audience or client.

Ideally a good human translator, on the other hand, is expected (unlike a translation machine which, as far as we know to this day, is – unfortunately – incapable of this) as a matter of principle to establish a symmetry between the two parties by (1) being equally at home in both languages, by (2) not being a partisan of either party, by (3) making an equal effort in both directions to understand the utterances of his clients, and by (4) checking that his own utterances are equally well understood on both sides. It is difficult to provide a functional definition of the capability of this idealised translator, which even in case of peak human performance is unlikely ever to be realised, since the translator would also need to be equally 'at home' not only in the two language cultures, but also in the styles of thought and speech and the interests of his two clients. The type 4 language machine thus faces the problem that what it is actually supposed to do (so far) remains underdetermined (if not entirely undetermined). We have only the formal criterion of equivalence between what the translation machine does and what the human translator does, but we are not able to say exactly what it is substantively that has to be the same, in such a way that would allow us to look for a technological strategy for getting machines to do it.

It goes without saying that – as with any substitution of human work – the ultimate aim is only equivalence of outcome, and not functional equivalence between human being and machine. This is illustrated by the example of a machine performing a calculation. The machine is required only to produce the same (correct) result that the well-versed human calculator must produce. But this is not to say that it must 'go about it in the same way' as a human being does. Similarly, a translation machine is required only to achieve the same outcome as an idealised human translator, but is not required to achieve it by following the same procedure. The procedure followed by the human translator will be linked to his language learning history, and perhaps his entire bilingual biography.

So should it prove to be the case that a 'semantically correct translation' of a text in a natural language can be produced, or at least approximated, by purely syntactic means, or in other words, should it be that the translation machine 'functions' differently to the human translator, this would not be a (sufficient) objection to the possibility of a machine performing the task of translation well.

As we search for a more precise definition of what it is that a human translator does, the focus shifts away from the human beings who are intentional agents, i.e. the translator and his two clients, and onto a property of languages. Although languages as a whole and any language utterance in itself (including those within longer bodies of text) do of course always remain a product of intentional action by human individuals, the success or failure of the actions of the translator may depend on properties of the language or languages that are invariant regardless of the speaker and listener (i.e. apply to any and all speakers, listeners and human translators). In other words the focus shifts onto the issue of whether, in order to design a good translation machine, the condition must be met that two texts in different natural languages must be 'translatable' in the sense that the texts themselves can be structurally related to each other. (The significance of this condition becomes clear when we consider the example of attempting to translate a set of operating instructions for a modern electronic device into a 'dead' ancient language, in which the most elementary words required would be lacking.) To this end texts, and indeed all linguistic utterances and the rules to which they are subject, will need to be stylised such that they apply 'to all speakers and listeners' of a language, and in the case of translation, to both natural languages.

It is clear from the above that singular texts such as a lyrical poem, in which the author consciously works with allegories, metaphors, irony, imprecise associations and other stylistic means, are less suitable for testing the quality of translation machines than texts on standardised practices. Examples of the latter are found in fields where understandings have been harmonised or even standardised interculturally in technological, bureaucratic, academic or other settings.

Yet for these examples too, the philosophical question as to the criteria for improving a translation machine remains difficult enough.

6. Suspended Intentionality

The idealised human translator is as it were a person with no intentions of his own as far as the tenor of the translated texts is concerned. His intentions are rather formal, in the sense that he is required merely to establish symmetry and an optimal translation in both directions. So in this sense we could say of an idealised human translator that his intentions concerning the actual tenor of the text are 'suspended'. As long as he is translating he does nothing but suspend his intentionality. He does not participate in the discussion between his two clients that he is translating.

This brief observation shows that intentions always remain with human beings and neither should nor can be delegated to machines, even when those machines substitute what humans do. The intentionality of all speech acts is 'suspended' – though not eliminated. Just as the machine providing timetable information by telephone does not pursue intentions of its own, but merely fulfils the purpose it has been designated by the railway company, namely to provide a customer-friendly information service as cheaply as possible, all machines, including translation machines, can be designed to serve their *purpose, which remains external to them* – in this case is to provide an optimal translation. This purpose is an external, overarching purpose that encompasses the internal, substantive sub-purposes that are localised in the intentions of the persons involved (translator and clients), and therefore naturally remain with their human authors. Nonetheless, a fundamentally new element does come into play here: *a new type of technology* (as promised by the title of this essay).

The fact that intentions are suspended in the design and use of machines does not prevent these machines from being given an *internal freedom of choice* concerning sub-ends and strategies selected in relation to their overarching ends (which remain external). Consider the (simpler) example of machines that do not use language such as autopilots for an aircraft or a lorry. They are programmed (using additional parameters such as speed, minimisation of fuel consumption etc.) to reach a target destination, but in case of a storm or a traffic jam on the motorway 'decide' 'autonomously' to change route. The anthropomorphic locution of 'autonomous decision-making' is unproblematic in the sense that both the human designer and the competent user know that the alternative options available need to be programmed not specifically and individually, but as 'leeway for action' based on an internal 'evaluation system'.

When using machines like this, it is crucially important for legal and liability reasons that *responsibility for a malfunction* leading to a plane crash or a collision can be clearly ascribed to a person. For each (legally permissible) case of use, it will be necessary to determine whether this is the pilot/the driver, or the designer, or the operator of the machine, or some other person. This is because only humans can perform actions and pursue intentions, and select means and achieve ends, for which they can be held responsible. It is a characteristic feature of the natural science agenda that machines, on the other hand, just like animals or plants, cannot be held responsible for damage or accidents. They merely 'function' in a way that can be causally explained.

One area in which this approach can be studied is chess computers. They are designed in a specific game situation to (1) go through the moves available to them, and hierarchise these according to the importance of the pieces, (2) generate trees of possible moves by the opponent, and then (3) develop further trees of moves available to themselves, which (4) they then evaluate in terms of gains and losses. All this of course is subordinated to the overarching end of winning the match (or at least achieving a draw). We do not need to reinvent the wheel here when we speak of 'suspended intentionality' in the sense of an overarching end or purpose of the machine that is unproblematically compatible with a freedom of choice for sub-ends and means. In the case of translation machines, 'suspended intentionality' refers to the external, overarching end of the optimal, symmetrical translation, which leaves scope for freedom of semantic choice as sub-ends of the two clients and their texts.

In the case of language machines, however, this structure is different again. Where the design of machines is based in a general way on knowledge that enables them to avoid and eliminate malfunction[11], as in the case of the autopilot and the chess computer, the scope of possible 'responses' is explicitly prescribed (for instance in the control parameters). It is the degrees of freedom of the aircraft or the lorry, or the degrees of freedom of the chess moves allowed by the rules, that define in advance what 'freedom of choice' exists for the aforementioned sub-ends and means. In the case of a translation the situation is – at second glance – fundamentally different.

If what a good human translator does is to be substituted (as well as possible) by a machine, then we need to remember that, broadly speaking, we are dealing with a *hermeneutic object*. Here there is no scope that is predetermined by parameters or rules of play that allow possible moves. This means that whether or not a malfunction such as an imprecise, misleading or incorrect translation occurs that *jeopardises the overarching end* of the 'good translation', *can only be decided on by a competent speaker and listener* exercising their judgement on the imprecise, misleading or incorrect translation. A merely internal understanding of the text based on syntactic and semantic criteria must for instance always be complemented by an adequate situational or contextual understanding on the part of a human being, relative to which the uncompromised nature of the good translation is judged.

No attempt will be made here to scale the heights of solving familiar longstanding problems of philosophical hermeneutics. We will aim rather to keep our feet on the ground as far as possible, and focus on

---

[11] In the works on methodical culturalism, the generation and control of scientific knowledge (and the definition of a methodical concept of a 'law of nature') is based in a general way on technical action and the elimination or avoidance of malfunction. The uncompromised functionality of measurement instruments for observational and experimental apparatuses is the paradigmatic end, relative to which malfunctions and their elimination lead to experience. These definitions were first provided in: Janich 1973.

specific problems such as those that arise when we compare the qualities of various speech recognition systems (dictation systems).

The comparison of the Nuance 'Dragon naturally speaking' system with the IBM 'Voice pro' system published in the technology section of the FAZ newspaper on 2 May 2009 is helpful in terms of comparing like with like. The comparison applies criteria (though these are not made explicit anywhere) that remain within the competence of the intelligent speaker and listener. A human being dictates one and the same text to both machines using the respective software applications, and then reads what the computer has made out of it in writing. If this comparison is made in relation to the *same* sound event 'dictation' and the written results of the two programmes, then the criteria of the comparison are unproblematic. Yet here too, it is for the intelligent reader to decide whether or not the printed text accurately reflects the spoken dictation.

When these systems are uncertain what to transcribe, they offer alternative options. Here, the aforementioned newspaper article identifies a serious difference between the compared systems. One of them (Dragon naturally speaking) almost always offers a useful ('appropriate') option (which the machine then executes when prompted), while the other almost never has a suitable option to offer. The article then compares further functions of the two dictation systems by applying lexical statistics (relative frequency of the occurrence of various words together). This criterion for comparative judgement remains unclear, however. Does one of the transcription machines here simply 'do' something functionally differently to the other one (thus making it better or worse than the other), or might lexical statistics be fundamentally misleading, because they fail to do justice to the plastic, creative and constructive nature of language and the intentions of individual usage?

This brings us closer to defining the 'hermeneutic object', which as 'suspended intentionality' cannot be defined in relation to a formal end, unlike interactive information services (which deliver accurate information on train connections), pocket calculators (which perform the calculation correctly), chess computers (which win the match) or dictation systems (which type the text correctly in relation to the theoretically well-pronounced dictation). The latter cases all have clear criteria for uncompromised functionality, including operationlisation, e.g. when the train passenger catches her train on time. A corresponding freedom from malfunction in the optimal translation process, on the other hand, does not exist. Language comprehension and generation is of a fundamentally different nature because it remains tied to intentionally acting human agents, even in forms of communication that remain invariant regardless of the speakers and listeners. This can be illustrated in everyday language as follows: a translation must capture what it is that one person wants of another when they speak to them. (And this is not decided by an observer located at an Archimedean vantage point outside of the communicative situation. In case of doubt, the two interlocutors themselves must reach a consensus on it.)

7. The Hermeneutic Specifications (Conclusions)

- For the substitution of language-based and directly language-dependent human action by machines, the only type of criteria available are based on uncompromised functionality, as judged by competent ('intelligent') language users in relation to specific situations.
- A form of gradual improvement can be achieved by moving from individual to inter-individual and trans-individual judgement of the 'correctness' of interpretation (i.e. translation) by language users. It is, however, illusory to believe that there can be any guarantee of convergence that will disambiguate language use. Nor can the gradual improvement of invariance vis-à-vis speakers and listeners compensate for the fundamental situational specificity (speaker, addressee, place time) of any linguistic utterance.
- A philosophical hermeneutic is required as a rational (scientific) method by which for instance *two competing interpretations (i.e. translations) of the same text* can be judged by explicit criteria. To date, no such science exists. Hermeneutic questions cannot (as things stand today) be answered by appealing to explicitly scientific criteria.[12]
- Therefore, the construct of suspended intentionality that is helpful in the case of other machines (overarching ends of the design, operation and use of cognitive machines producing

---

[12] Concerning the state of the art of rational hermeneutics, see Janich 2008: 371–381.

equivalent outcomes) with (metaphorical) internal autonomy within spaces of possibility predetermined by the designer cannot (to date) be transferred to hermeneutic machines. This is because the seemingly clear overarching end of the 'good' or 'optimal translation' is underdetermined, and can only be operationalised by appealing to intelligent human speakers and listeners.

- This, however, means that 'what human beings do' has not been 'substituted' by a language-using machine. The human being as an agent of judgement remains indispensable, also with respect to the criteria sought for the uncompromised functionality of a machine.
- One pragmatic way out of this might be a regionalisation of texts by areas and practices (shared by the clients and the human translator). The competent judgement of translations by human beings might then generate feedback processes for the selection of criteria to distinguish better translations from poorer ones. The linguistically competent translation critic would say how the translation should be improved; the software engineer would attempt to generalise this information and form criteria, whose application in specific cases would then again need to be judged by the same competent speakers.
- The *innovative aspect of this new technology* is that *no fixed criteria for uncompromised functionality* can be formulated for the design of language machines. All that can be formulated is a list of 'hermeneutic specifications' designed to support ongoing change to improve the processing of language. These specifications cannot define (or operationalise) the overarching purpose of a language machine as being of the uncompromised functionality type.
- Metaphorically speaking we could describe the 'hermeneutic object' of language machines by saying that the only way to bring their performance into line with what human language users do is to get the machines to 'learn' from human performance. Here, though, equivalence of outcome as an end and avoidance of malfunction as a means have become something different from what is found in all other machines.

References

Gutmann, M. (1996): *Die Evolutionstheorie und ihr Gegenstand. Beitrag der Methodischen Philosophie zu einer konstruktiven Theorie der Evolution*, VWB, Berlin.

Janich, P. (1973): "Zweck und Methode der Physik aus philosophischer Sicht", *Konstanzer Universitätsreden*, edited by Gerhard Hess, no. 65.

Janich, P. (1993): "Das Leib-Seele-Problem als Methodenproblem der Naturwissenschaften", in: Elepfandt, A. & Wolters, G. (eds.), *Denkmaschinen? Interdisziplinäre Perspektiven zum Thema Gehirn und Geist,* Universitätsverlag Konstanz, Konstanz.

Janich, P. (1993): "Mensch und Automat. Philosophische Überlegungen zur technischen Substituierbarkeit des Menschen", in: *Bemannte Raumfahrt im Widerstreit*, Köln, pp. 25–34.

Janich, P & Weingarten, M. (1999): *Wissenschaftstheorie der Biologie,* Fink Verlag, München.

Janich, P. (2000): *Was ist Wahrheit? Eine philosophische Einführung*, C.H. Beck, München.

Janich, P. (2001): *Logisch-pragmatische Propädeutik,* Velbrück, Weilerswist.

Janich, P. (2006): *Was ist Information? Kritik einer Legende,* Suhrkamp, Frankfurt a. M.

Janich, P. (2008): "Hermeneutik und Rekonstruktion. Probleme einer Philosophie des Exakten"*,* in: Bernhard, P. & Peckhaus V., *Methodisches Denken im Kontext. Festschrift für Christian Thiel,* mentis, Paderborn, pp. 371–381.

Janich, P. (2010): *Der Mensch und andere Tiere. Das zweideutige Erbe Darwins,* edition unseld, Berlin.

Action and Autonomy:
A Hidden Dilemma in Artificial Autonomous Systems

Mathias Gutmann, Benjamin Rathgeber & Tareq Syed

Abstract: Within the context of roboethics, the possibility of morally acting autonomous systems is often proposed; this kind of systems seems to provide unexpected threats for the thesis that only humans can act morally or evaluate their actions ethically. It remains uncertain so far whether such systems are possible at all or whether there are principial, non-empirical obstacles, which might contradict even their pure conceivability. This paper examines the structure of attributing "action" and "autonomy" to technical systems. Some central characteristics of autonomy are reconstructed and their systematic relationship to "action" is identified. It can be shown that an immanent contradiction is generated when we assume that artificial systems are capable of acting autonomously.
Keywords: autonomous systems, strong-, weak-, negative-, positive-autonomy, roboethics

0. Setting the Stage: Roboethics

On the first sight roboethics seems to be a new branch of applied ethics, such as "bio-", "eco-" or "genethics", which deals with technical engineering, informatics and robotics. From this perspective roboethics marks the emergence of a new kind of moral, ethical or societal problems and at the same time the development and differentiation of a philosophical research program, which deals exactly with the respective problems. Roboethics shares some features with other types of applied ethics, as for example relevant problems are often reflected in advance; that is, the techniques in questions are only in an experimental state or the object of pure thought-experiments.

However, the already existing autonomous systems are usually more or less sophisticated extensions and optimisations of (normatively) trivial systems such as factory robots, servo-systems of different kinds and driving-assistance-systems. As such, they provide not much more than classical legal and ethical problems, which are usually connected with the reliability of the respective technique, the legal responsibility for the production and the application of those systems. These are problems, which are covered by ELSI-Projects (s. Decker et al. 2011). Of a more challenging type are technical systems, which are not (fully) implemented yet or in a state of planning such as "ambient intelligence", "ubiquitous-" and "organic computing" or "autonomously acting robots". The ethical and societal dimensions of those systems may provide a considerable threat not only to societal structures, but particularly to the self-understanding of modern human beings. If certain aspects of personality-like autonomy (e.g. the freedom of decision and choice) were effectively implemented into purely technical systems, then "being human" would only *contingently* be connected with individualised, biotic entities. From this point of view, even one of the presumably specific human features, namely the embodiment of knowledge and know-how, seems to provide no insurmountable obstacles for technical simulation.

Accordingly, the most radical approach would have to deal with robots or other technical systems not as pure objects but as subjects of ethical assessment. They would be subjects in the sense of ethically evaluating their own actions, or acting in accordance to moral rules – and consequently they were not just a pure extension of existing autonomous systems but a fundamental new type.

1. Technical Systems as Subjects?

In order to characterise the problems generated by dealing with autonomous artificial systems in the role of moral agents, let us refer to Asaro's (2006) assumptions on robots as authors of ethical judgement. Asaro draws a strict difference between those artificial systems, which may be technically sophisticated but nevertheless display no specific moral complexity, and systems which are morally complex. The first group is represented by driving systems, medical assisting systems like surgical manipulators etc. Their moral simplicity is defined in one central respect:

> "A driving system ought to be designed to obey traffic laws, and presumably those laws have been written so as not to come into direct conflict with one another. If the system's actions came into conflict with other laws that lie outside of the task domain and knowledge base of the system, *e.g.* a law against transporting a fugitive across state lines, we would still consider such actions as lying outside its sphere of responsibility and we would not hold the robot responsible for violating such laws. Nor would we hold it responsible for violating patent laws, even if it contained components that violated patents. In such cases the responsibility extends beyond the immediate technical system to the designers,

> manufacturers, and users – it is a socio-technical system. It is primarily the people and the actions they take with respect to the technology that are ascribed legal responsibility." (Asaro 2006: 13)

Following Asaros argument, it is primarily the creator of the system – in some lesser degree even the user – who is held responsible for the actions of the system. The rules which the system is described to follow, resemble the relevant respective laws to a certain degree, which exceed not necessarily the level of functional, technical rules.[1]
In this case technical systems are only in a weak descriptive sense *autonomous*, their freedom of action allows us to understand them as pure objects of moral sentiment. The second kind of systems then is *autonomous* in a morally very demanding way:

> "Real moral complexity comes from trying to resolve moral dilemmas – choices in which different perspectives on a situation would endorse making different decisions. Classic cases involve sacrificing one person to save ten people, choosing self-sacrifice for a better overall common good, and situations in which following a moral principle leads to obvious negative short-term consequences." (Asaro 2006: 13)

These main criteria are given in terms of utilitarianism here,[2] but we have to observe only whether the structure of this argument is convincing. The methodologically relevant point is that technical systems are supposed to be *built* in order to evaluate their own actions in terms of moral sentiment. Even if an ethical *Turing test* for robots seems to overstretch the point, Asaro emphasises the necessity to design artificial agents, which not only act according to laws but which are able to moral evaluation of their actions. The systematic implications of these systems can be identified by following an

---

[1] "Rule" is used in broader sense here, as is "to obey". We will see further below, what exactly is meant with intentional expressions of this kind in case of non-human beings. This is exactly the reason why we preferred the term „to follow" at this state of argumentation (nevertheless, the problem of intentional expressions remains unsolved so far).

[2] This restriction is of some relevance, insofar the focus on utilitarianism reveals a blind spot of roboethics: by reducing ethical reasoning on utilitarianism, the problem of evaluation and judgement becomes concealed. In order to provide the respective adequate pay-off-matrices and optimisation rules we tend to neglect the fact that even strong utilitarian thinking necessarily presupposes some general rules, which cannot be justified by utilitarianism as such (s. Mill 1974, Tugendhat 1995).

escalation strategy, starting for example with driving systems and finally approaching the realm of autonomous weapon robots:

> "Rather, we should seek out real-world moral problems in limited task-domains. As engineers begin to build ethics into robots, it seems more likely that this will be due to a real or perceived need which manifests itself in social pressures to do so. And it will involve systems which will do moral reasoning only in a limited task domain. The most demanding scenarios for thinking about robot ethics, I believe, lie in the development of more sophisticated autonomous weapons systems, both because of the ethical complexity of the issue, and the speed with which such robots are approaching. The most useful framework to begin thinking about ethics in robots is probably legal liability, rather than human moral theory – both because of its practical applicability, and because of its ability to deal with quasi-moral agents, distributed responsibility in socio-technical systems, and thus the transition of robots towards greater legal and moral responsibility." (Asaro 2006: 15)

In this case, robots would appear to be moral agents and as such real subjects – not only objects of moral sentiment. This proposal raises a whole bunch of philosophical questions; nevertheless, according to the methodological aspects we will focus on two points, which allow us to specify semantic and pragmatic objections against the possibility of building morally acting systems. The first argument is connected with the methodological structure of referring to something as an autonomous system, the second with the logical grammar of human action. The main task of the following pages is to show that both questions are finally an implication of the phrase "to built a morally acting system[3]". In this context we will start with the question what is meant by the term "acting", which nevertheless displays the methodological shortcomings of artificial agents in a nutshell.

2. Conceptual Distinctions

Whatever might be implied in terms of etymology, "actions" can be criteriologically defined as means, which are applied to achieve specified ends (s. Hartmann 1993 & Janich 2001). This statement can

---

[3] The term "system" is used here in the weakest sense, referring to any technical artefact, which is used in the framework of human action; it simply discerns artificial from natural systems (concerning the methodological relation between artificial and natural systems; s. Gutmann 2010, Gutmann et. al. 2011).

be understood in two different ways: firstly, as a strong disjunction between "actions" and something like "behaviour"; secondly, as a hermeneutic pretext (discussed in chapter 2.3 and 2.4). The first distinction refers to strong differences insofar as what is described as action cannot be understood as behaviour and vice versa. The second distinction then understands action and behaviour as aspectual differentiation referring to human activities in a strong description.

2.1   How to Act?

Following the first strategy, we have to assume that "behaviour" and "action" are defined as contrast-classes. Accordingly, some non-overlapping elements can be identified (s. Janich 1997 & 2001):

1. *Behaviour* is governed by natural laws, that is, the performance of a specific behaviour is not arbitrary or deliberate. The question – for why there was a specific behaviour – is answered in terms of causes; and consequently, it is rather senseless (by definition) to ask somebody to perform a behavioural movement (the iris- or the patellar-reflex may serve as examples as well as some hiccup). From this perspective neither the effects of the respective behaviour nor relevant side-effects are ascribed to the person: these instantiations of behaviour are in the sense of the word "accidents". The logical grammar of *behaviour* is that of effects, generated by causes (which may be effects of further causes again and so on). The answer then, why a specific behaviour took place, follows the scheme of an explanation (e.g. covering law explanation; s. Hartmann 1993; Hempel 1977).
2. Accordingly and in contrast to behaviour, *action* then is defined as a voluntary activity, which is governed by *reasons* and not by *causes*. Due to the practical nature of actions their adequacy is evaluated in terms of a practical syllogism, whose antecedent contains the instructions. These instructions are assumed to be relevant in reference to the respective context, defined by the adequacy of the application of specified means due to given ends (s. v. Wright 1971). The logical grammar of action is that of results generated under the guidance of explicit and explicable purposes.

Comparing *action* to *behaviour* in the light of the criteria above, we can determine the relevant similarities and differences most clearly. Both are objects of hypothetical constructions but the status of the construction differs fundamentally:

1. In case of behaviour, the "if"-term contains all conditions, which are relevant for the subsumption of the given experimental situation as an instantiation of a specific type (named S1). The law term then contains the correlation between S1 and a resulting situation of a second type (named S2); the explanandum term finally contains the specific S2 situation, which results from the given S1. Consequently, S2 is supposed to emerge without further intervention from the experimenter. The experiment then is a kind of a machine, which produces correlations of a specific form (s. Janich 1997). When working with living entities, their behaviour would be described as a "natural process" which eventually happens even without the intervention of the experimenter[4]. However, the preparation, initiation and conduct of the experiment, including the registration, processing and interpretation of the data produced are *actions* of the experimenter. Whereas the natural behaviour (e.g. of an animal) is observed, the activity of the experimenter provides a perfect example for action.
2. In the case of action, the law term explicates the means-end-relation; so, if a certain purpose is to be achieved, certain means that are known to be adequate, ought to be applied (Wright 1971: 93ff; for further reading on the practical syllogism s. Anscombe 1957).
3. Practical relations then between ends and means are not of a causal nature, whereas causal relations can play a crucial part within the means term; i.e. causal relations may serve as means of a specific type. Nevertheless, they only do so in accordance to the ruling ends-means-relation (s. Dewey 1925, Sellars 1963, Gethmann 1979, Lorenzen 1987, Tetens 1987, Janich 1997).

If we define action in terms of means-end-relationships (whose nature we have to consider in greater detail later), the "deliberateness" is

---

[4] It is so far only some kind of Skinner-Box experiment we used as a standard; but the form of "experimentalist action" is the same even in terms of physical, chemical or geological research (for further reading s. Hacking 1996, Janich 1996).

given by reference toward the respective ends[5]: it is only the determination of a specified end that implies the necessity of actions of a specified kind. And even this necessity is relative, as it must be referred to the successful application of (specified and specifiable) means. However, discerning behaviour from action on the basis of this reconstruction we are provided with some criteria that characterise action (Janich 2001):

1. The deliberative structure implies that it is – and always must be – possible to freely determine the ends, which are supposed to be realised by acting. Consequently, an action can be ascribed to a person as the use of means in order to realise determined and determinable purposes. Behaviour then is not ascribable in the same way, insofar as no ends-means-relationship is underlying.
2. The ascriptive structure of action implies the responsibility for the action and its immediate results (as well as for possible side-effects).
3. The practical structure of action implies the (normative) capabilities and skills of a person, whose acts have to be justified (rationally) in reference to the underlying purposes.

These criteria are strict insofar as we would not assume that behaviour realises ends. Dealing with ends in terms of natural systems or entities (such as animals or plants structuralised as organisms[6]), "end" is a metaphor, which can be replaced by stating an "as if relation" with a model construction following.

From this point of view, it is clear by definition that only persons can act, and that persons are persons by nature. This however may result in some severe counterarguments, which are based upon the assumption that it might be possible to implement technical systems some abilities of rational reasoning. In this case it would be the *factum brutum*, that – till now – rational reasoning is connected with some contingent results of biological evolution: namely the fact, that it is

---

[5] This situation changes fundamentally if we refer to more hermeneutic concepts of action (s. 2.3 and 2.4).
[6] It is important to realise the difference between living entities and organisms (Gutmann 2002). Organisms are the result of the s.c. weak ascriptive procedure, which starts with the (metaphorical) description of living entities (i.e. elements of everyday lifeworld practice) as "if they were" functional units (e.g. engines, lever-constructions or other types of machines).

Homo sapiens who addresses exclusively himself as wise and understanding.

2.2 Acting Autonomously by Nature?

Discerning behaviour from action by referring the first to causal, the latter to pragmatic relations, *autonomy* can be attributed to both of them. Consequently, the term "autonomy" becomes rather ambiguous, as it can be an element of object- as well as meta-language-level. In the first sense it simply attributes processes, which are assumed to be instantiated without external interference; Saturn revolves around the sun, plants grow in the direction of the sunlight, and animals show specific locomotive patterns. But all these processes just happen, and are insofar "autonomous". This weak sense of autonomy contrasts significantly to the origin of the term. As Sturma (2003) points out, "autonomy" is by origin a political concept, dating back to antic time, during which autonomy refers e.g. to a (city-)state concerning its legislation in the full sense, whereas heteronomy implies the dependency on the legislation of some other (hegemonial) political body. In more recent approaches, autonomy is constantly connected with personality, a connection Sturma explicates by avoiding the metaphysical problems of "will" and "freedom of will". In this case, autonomy becomes a feature of a person. Autonomy then is "logically" connected with the ability of a person to act within the "realm of reasoning", a connection that presupposes the strict semantic difference between *reasoning* and *causation*. Whereas causation belongs exclusively towards the language games of sciences, the term "reason" is supposed to belong to a richer language game, which includes the argumentative ability to justify actions.[7] This

---

[7] Sturma (2003: 42): "Gründe sind Bewertungen und – in der Form von Handlungsgründen – Veränderungen von Sachverhalten. Sie erfüllen überdies Funktionen in Erklärungen und Rechtfertigungen. Der Raum der Gründe ist dementsprechend der begriffliche Rahmen für den systematischen Zusammenhang von Begriffen sowie Regeln und Relationen von Begriffen und Aussagen. Personen verwenden im Raum der Gründe Ausdrücke und Aussagen, mit denen sie schließen, begründen, rechtfertigen und ihre Praxis an den eigenen Erwartungen sowie an den Erwartungen anderer Personen ausrichten. Innerhalb des Systems des Raumes der Gründe können Personen einerseits Aussagen über die Einstellungen und das Verhalten von Personen, andererseits Aussagen über Ereignisse in der Welt machen."

argumentative ability is connected to the status of the arguing parties as persons; and expresses itself by the autonomy of acting within the "space of reasons". At least insofar as the participation in a lifeform (in the sense of *bios*; s. Gutmann 2011) is a necessary condition of acting humanely, the difference to automated systems becomes clearcut even if we consider these systems to be able to act within the "space of reasons". They do so only insofar as they follow rules, but they do not insofar as they are not able to creatively deal with notions. Consequently, the use and understanding of metaphors and analogies is one of the most important criteria for specific human creativity. This criterion is itself finally based upon a "self-referential and semantically justified" use of language. However, insofar as humans are by nature destined to generate personality, being an autonomously acting agent becomes coextensive to being a specimen of *H. sapiens*.[8] The methodologically most interesting point here is that the difference between technical and natural systems (of which *H. sapiens* is just one example) is based upon the "fact of reason", that *H. sapiens* is capable of reasoning. This difference would collapse in the very moment in which cognitive expressions like "intelligence" were attributed to certain technical systems. Sturma rejects a structural analogy between the attribution of those expressions to technical and natural systems on grounds of the latter's lack of intentionality.[9]

---

[8] "Der Raum der Gründe ist Ausdruck des folgenreichen Sachverhalts, dass Menschen über einen langen Zeitraum ihre Naturgeschichte in Kulturgeschichte transformiert haben. In ihm sedimentiert sich formal und inhaltlich die Entwicklung der Begriffe, die den epistemischen, moralischen und ästhetischen Eigensinn menschlicher Fähigkeiten und Eigenschaften konstituiert sowie einen neuen Rahmen für Denk- und Verhaltensweisen erzeugt. Semantik ist geronnene Kulturgeschichte. Deshalb kann auch davon gesprochen werden, dass mit dem Raum der Gründe eine neue Lebensform hervortritt." (Sturma 2003: 43).

[9] "Von Intelligenz kann im nicht-übertragenen Sinne nur unter der Voraussetzung intentionaler Bezugnahme bzw. einer Semantik mentaler Repräsentationen gesprochen werden. Das Bewusstsein von Personen ist durch sein komplexes und kompliziertes System von Bezugnahmen definiert. Dabei kann ein und dasselbe Ereignis zum Gegenstand unterschiedlichster intentionaler Beziehungen gemacht werden: Eine Person kann einem Ereignis erwartungsfroh entgegensehen und gleichzeitig in Sorge sein, ob die Erwartungen berechtigt oder etwaige Hoffnungen überzogen sind. Ein derartiges Einstellungsgeflecht gehört zu den charakteristischen Ausdrucksformen menschlicher Intelligenz. Seine wesentlichen Strukturmerkmale sind Intentionalität sowie syntaktische und semantische Präsenz im Raum der Gründe." (Sturma 2003: 46).

Consequently, robotics seem to have no severe normative implications insofar as robots are to be understood as pure tools only (to given human ends). However, the problem remains whether the criterion of rational argumentation is adequate in order to substantiate the differentiation between natural and technical systems. The very moment we were forced to admit that a technical system passes the Turing test in regard of rational argumentation, the only difference which could be stated is the difference between a (technically) produced and a (naturally) generated system. This difference is normatively irrelevant as long as we are not able to demonstrate how rationality depends on the status of something as a naturally generated unit. If not, Sturma's conclusion is inevitable, that in case artificial persons were actually built, they would have to be acknowledged as full-fledged persons.[10] According to Sturma, the only reason which prevents us from building artificial persons would be a negative answer on his question, whether we should be allowed to build artificial persons.

However important we consider the status of a person to be the ultimate keystone of a strong concept of autonomy, the methodologically relevant question remains whether we are reduced to more or less technical or empirical objections. Following Sturma's line of argument, we had to state that up to now we are just not capable of (technically) producing artificial agents, which are sufficiently similar to "biologically" generated "natural" persons. In the following step, we have to scrutinise the methodological structure of the term "autonomy": This allows us to gain a basis for the consideration whether or not we might have to assume fundamental non-empirical

---

[10] "Falls es aber einmal möglich sein sollte, Roboter zu entwickeln, die im Raum der Gründe selbständig agieren und über Selbstbewusstsein und mentale Repräsentationen verfügen können, müssten wir sie in unsere ethische Gemeinschaft aufnehmen. Wir dürften sie nicht länger wie Maschinen oder technische Sklaven behandeln. Dann sollten wir uns aber auch mit der existentiellen Situation künstlicher Personen beschäftigen. Auf die Frage einer künstlichen Person, warum wir sie in maschineller Form überhaupt zur Existenz gebracht hätten, wären wir kaum besser vorbereitet als der unglückliche Dr. Frankenstein. Wenn aber ernsthaft Projekte der künstlichen Erzeugung von Bewusstsein erwogen werden sollen, dann wäre es ratsam zu fragen, ob es überhaupt rechtfertigungsfähige Gründe dafür geben kann, auf technologischem Wege neue Bewusstseinsformen mit existenziellen und ethischen Eigenschaften zu entwickeln." (Sturma 2003: 52)

objections against the ascription of autonomous action to artificial systems.

2.3 How to Act Autonomously I: Negative and Positive Autonomy

Our argument so far was developed without any considerable reference to ethics or morality. The reason for this restriction results from the structure of the argument, which refers to actions as means (applied to achieve ends). In regard to this practical structure of our starting point, we can deal with actions in terms of technical relations, that neither presuppose ethical premises nor imply ethical consequences, whereas ethical reasoning is – at least to certain extent – tightly connected with actions. A very similar asymmetry seems to rule the relation between morality and autonomy: even if autonomy is considered to be a necessary prerequisite of morality, the converse relation is not necessary; accordingly, we can deal with autonomy without implying strong ethical consequences.

The term "autonomy" then is analysed best by referring to its immanent semantics on the one hand and to its counterpart (heteronomy) on the other. By origin being a political concept it shows an inherent normative structure[11]. As such, it cannot be identified on descriptive grounds alone and nevertheless certain descriptively controllable criteria are to be observed in case of stating that "something shows autonomy". The meaning of autonomy is explicable by replacing the noun with an attribute that serves as an adverbial or adjective qualifier. Accordingly, we would have to deal not with the opaqueness of autonomy but with "autonomous x" or the "x performing y autonomously". By avoiding the noun in the first place, we gain an alternative approach, because here, "autonomous" became an expression that qualifies human action or activities; the term "autonomy" then becomes a (at least) three-termed expression, with which we state that "A is doing x autonomously in reference to y". Stating that an action x is autonomous implies that in order to do x,

---

[11] A normative structure does not imply ethical relevance. Technical norms are without any doubt norms, but they are ruled by a practical syllogism, which considers the adequacy of means according to given ends. Industrial norms may serve as an excellent example here; the end however has to be justified; and this justification cannot be done by referring to pure technical know-how.

the actor (A) does not face obstacles that hinder A to perform the action. What could be meant with "obstacles" becomes clearer when we take a closer look to the immanent structure of means-end-relationships. By stating that actions are means in order to achieve purposes, we are confronted with the fact that we have to refer to the ends twice:

1. Ends are given, they are "ends-in-view" (s. Dewey 1925) – and in this sense they have a propositional aspect, which represents a counterfactual state of affairs. The description of this counterfactual is necessarily accompanied by a performator, which demands for the realisation of the counterfactual state of affairs.
2. Ends are realised by applying means. In this sense the realised ends provide the criteria for the successful instantiation of the application of means.

Whether or not the purpose of an action is achieved depends not only on the successful performance of the application of means but on further conditions, which are neither logically nor conventionally connected with the ends to be realised. It is far more the experimental character of actions, which constitutes the relation between the "end-in-view" (Dewey 1925) and the "realised end". The constitution of this relation is the result of the action itself. "Action" then becomes a way of generating relations between *ends* and *realised ends*; and insofar the term "action" itself becomes ambiguous. In one sense, "action" simply designates what has to be done in order to achieve x, i.e. to realise an end-in-view. In a second sense, "action" designates the way, in which the relation between the end-in-view and the realised end is generated.

The final achievement of an action then is not simply implied – neither by the successful performance of the action schema nor by the availability of the relevant resources (which are elements of the antecedent of a practical syllogism). The final achievement of an action, i.e. the ultimate fulfilment of a purpose, is but the subject-matter of a contextual deliberation, the results of which cannot be anticipated – except possibly in the cause of a highly standardised and regulated context[12]. The discourse about the acknowledgement of the

---

[12] Such contexts are provided by scientific practices, e.g. in terms of experimental action. But even in these cases the problem remains to determine exactly which

actual achievement is a constituting element of what might be called a "successful action"[13]. We can now unfold the immanent structure of means and ends, and their connection with action allows us to discern positive from negative autonomy:

1. Referring to the given end-in-view and their realisation by applying certain means, negative autonomy is achieved if and only if the realisation of the ends-in-view depends upon the application of the means alone. The antecedent of this practical syllogism will be complex of course, as it has to cover a set of factors and conditions, which are to be considered. However, the realisation of ends is achievable even in cases of complex ends; that is, if the "end" can be analysed as a compound of sub-ends. The following hierarchy of sub-ends is then to be considered as the result of iterated realisation processes and can be descriptively subsumed under the antecedent.
2. Positive autonomy refers to the deliberative process of negotiating the relation between ends and realised ends. This process is structured communicatively and implies the acceptance of certain rules that constitute the discursive framework (s. e.g. Habermas 1992).

In the light of this differentiation positive autonomy must not be identified with "setting ends" or "determining purposes". The final achievement of the tool-application follows neither logically nor causally but from its respective performance: the setting of ends is but a moment of the action itself and its performance.
The performance of an action can be hindered by certain obstacles, which are derivable from the description of the respective tool-application. Following Heidegger (1962), there are – at least – three relevant types of obstacles, which are to be considered as a disturbance of actual tool-applications:

1. *Conspicuousness* ("*Auffälligkeit*") is generated by the disturbance of the "ready-to-handedness" of equipment: "We discover its unusability, however, not by looking at it [i.e. the equipment, the

---

context is relevant for a given scientific description, providing the antecedent of an explanation.
[13] In consequence, "successful" is not a primitive, monadic term.

authors], and estabilishing its properties, but rather by the circumspection of the dealings in which we use it. When its unusability is thus discovered, equipment becomes conspicuous. This *conspicuousness* presents the ready-to-hand equipment as in a certain un-readiness-to-hand." (Heidegger 1962: 102f)

2. *Obtrusiveness* ("*Aufdringlichkeit*"): During tool application some tools may be identified as being needed but not to hand at all: "Again, to miss something in this way amounts to coming across something un-ready-to-hand. When we notice what is un-ready-to-hand, that which is ready-to-hand enters the mode of *obtrusiveness*. The more urgently [Je dringlicher] we need what is missing, and the more authentically it is encountered in its un-readiness-to-hand, all the more obtrusive [um so aufdringlicher] does that which is ready-to-hand become – so much so, indeed, that it seems to lose its character of readiness-to-hand." (Heidegger 1962: 103) The obtrusiveness is defined only in terms of contextuality, as it is the availability of tools, which defines what is missing.

3. *Obstinacy* ("*Aufsässigkeit*"): If we are confronted by tool-application neither with the unreadiness nor with the missing of something, the achievement of our purposes can be hindered nevertheless, if "something stands in the way": "That to which our concern refuses to turn, that for which it has 'no time', is something *un*-ready-to-hand in the manner of what does not belong here, of what has not as yet been attended to. Anything which is unready-to-hand in this way is disturbing to us, and enables us to see the *obstinacy* of that with which we must concern ourselves in the first instance before we do anything else." (Heidegger 1962: 103) Obstinacy can be represented in different ways – as *obtrusiveness* and *conspicuousness*, too. It can be an unforeseen event, that prevents us from successful action as well as a state of affairs that has to be generated before our actual action can be performed etc.

The main point of this short Heideggerian reconsideration is the fact that "autonomous action" can be explicated in terms of negative as well as positive autonomy without necessary reference to strong scientific knowledge – at least for a methodological starting point. "Autonomy" then refers to rule-guided, purposeful action, which alone allows us to evaluate a given performance as an instantiation of an action of a specific type[14]. In this form "autonomy" does not contradict obligation but shows intrinsic connections to the form of obligation. Referring to Heidegger's reconstruction of tool-mediated action, it is the purpose which allows the differentiation between both central aspects of autonomy. Emphasising the tool-application, autonomy can be characterised as negative if there are no obstacles, which prevent us from achieving the determined purposes. Positive autonomy then emphasises the moment of the purpose itself and the process of determining the purpose.

2.4 How to Act Autonomously II: Weak and Strong Autonomy

From the reconstrution given above, we can assume that autonomy is not just a property of action at all but an expression on the meta-level, referring to something else.[15] For the sake of the argument, we can construct a two-dimensional matrix consisting of *autonomy* on the one hand and *heteronomy* on the other hand; thus we have four possible formal combinations, with two homogeneous (autonomous *autonomy* and heteronomous *heteronomy*) and two heterogeneous (*heteronomous* autonomy and *autonomous* heteronomy). From a conceptual point of view this distinction resembles Hegel's reconstruction of "acknowledgement", which explicates the use of "master" and "servant", starting with both as primitive (one place) expressions.[16] The contradictions explicated by the reconstruction of acknowledgement can be avoided, when the terms are in fact understood as relata of an – at least – ternary relation. One important consequence of this reconstruction is that "being master" does not formally exclude "being servant", nor does "not being servant" imply formally "being master". The relation however between both poles of the relation ("being master" and "being servant") can be explicated on different

---

[14] This restriction is of some methodological importance insofar as the construction of the term "action" necessarily refers to well-described and well-performed action-topics. And these topics are gained with regarding human activity as production and generation of something e.g. of shoes. Thus, we already have to know what is meant with "making shoes" in order to introduce the "action of shoe making", which in fact is an analytically (!) complex term.

[15] I.e., what we characterised as the realisation of ends by applying means and the deliberation of this achievement.

[16] S. Hegel 1986.

grounds (e.g. in terms of work, reproduction and laws; s. Gutmann 2004a). Applying this insight to our reconstruction of autonomy, we easily assume that "being autonomous" is not a primitive (one placed) expression but a multi-placed character:

1. Assuming the given possibility of setting ends that can be realised by applying means, we refer to *autonomous* heteronomy. The actor is free in setting his ends – however, he acts autonomously only insofar *as he sets ends*, which are necessarily determined to be achievable by defined means. Insofar, the author's action shows the character of strong autonomy, which should not be confounded with unrestrictedness.
2. On the other hand and at the same time, we refer to *heteronomous* autonomy by describing the process of setting the ends; here the actor is autonomous *in choosing between ends*. However, he is bound to the structure of the means, which are to be applied in order to realise the ends. Insofar, the author's action shows the character of weak autonomy, which should not be confounded with pure determinacy.
3. By referring to the pure act of setting ends – i.e. the pure ability to determine the use of something as a tool in order to realise ends – we refer to autonomous autonomy. In this description, the author of the ends acts as an author *per se* – an idealised, counterfactual construction, which can be explained only in reference to the relation of his acting *autonomously heteronomous* and *heteronomously autonomous*.
4. By referring to the pure realisation of ends – i.e. the pure potential of realising ends, a pure "in order to" – we refer to *heteronomous heteronomy*. In this description, the actor applies something as a pure tool, which then is referred to as a tool *per se*. Again, this seems to be an idealised counterfactual construction, which is accessible only by referring to the relation between acting *autonomously heteronomous* and *heteronomously autonomous*.

In order to understand artefacts acting autonomously, we refer to the "actor axis"; i.e. by stating that an artefact is realising ends autonomously, we state its acting as a heteronomously autonomous system. The ends are given by an actor, and these ends serve as criteria for the evaluation of the artefact's function. However, the determination of the way of realising the given ends is part of the function of the system itself – which may be considered a black box, as in the case of "neural networks". The actor, applying the tool however, is acting autonomously heteronomous as he sets the ends but attributes the realisation to the artefact, without further interaction. Being "autonomous" in this case means acting autonomously and this qualification does not exclude the guidance of rules, which govern the respective action[17].

We can now define s.c. "autonomous systems" as displaying weak, persons[18] as displaying strong and weak autonomy; consequently, if autonomous systems are described as acting in the role of a person, their functioning is described "as if" it were the acting of a person. They act in the role of persons only insofar, as the role-taking of a person is attributed to them. Scrutinising the replacement of a person's actions by the functioning of an artefact, we should observe that it is not the action, which is replaced, but the "action" described in terms of "operation". Thus, it is not *action* that becomes replaced but *human action, described as* the operation of an artificial system, producing functional states. The equivalence between the states of the system and the operations, which result from the operational description of actions, is possible only by reference to the purpose of the original action. This determination of the functionality of an artificial system remains adequate even if the ends are realised via steps which are only determinable by their outcome and not by the specific single steps. For example, if neural networks are used, which may be described as black-boxes considering the internal states of the net itself, the outcome has to be functionally equivalent to the determined ends.[19]

3. With Habermas against Artificial Autonomy

In a final step, we can develop an argument, which was directed originally against liberal eugenics by Habermas. By rearranging this

---

[17] If we assumed an exclusion of this kind, we had to defend some identity between "autonomy" and "indeterminacy", which is not just contra-intuitive but seems to fail the aim of proof: "indeterminacy" is a multi-termed expression, which can be explicated as a phrase of the following kind: "something is undetermined in reference to some knowledge A", with "A" representing a rule- or law-term.
[18] Which acts autonomously in terms of negative and positive autonomy.
[19] This is true even in the case that the final end is realised via the realisation of sub-ends; the compositionality is evaluated with respect of the final end only, which – as all ends – is set by persons.

argument in the light of our clarification of autonomy, we gain a structural argument against the morality of artificial persons, which is structurally independent from a naturalist foundation. We can even overcome the limits of Sturma's consideration, which at least opened the perspective of artificial personality in principle.

Habermas develops his core argument against the acceptability of liberal eugenics on the basis of (till now) utopian genetic skills to literally determine the phenotype of a child on ground of its genotype. That is, setting aside some counterarguments based upon epigenetic concepts, Habermas accepts the strongest possible premise of genetic utopia in terms of a thought-experiment. Given, then, the possibility to determine the phenotype of a child completely by determining its genotype, the parents could in fact decide about the "design" of their child. This thought-experiment transcends the narrow limits of recent genetic engineering, PID etc. by far, because we have to assume the actual capability to design the phenotype of a single person on grounds of genetic intervention. The resulting problems are not just those connected with prevention of diseases which may have an indisputable probability of being consensual; it is but the capacity to explicitely determine the genetic fate of the offspring itself in terms of the parent's design. Against the application of procedures of this sort Habermas (2001) presents some arguments, of which we summarise the three most relevant for our given context:

1. An anthropological reflexion based upon Plessner's (1975) differentiation between "being a proper body" and "having a corporeal body" ("Leib-Sein" versus "Körper-Haben"). However this seems to be a less convincing argumentative strategy as it presupposes a specific "biological" description of the human-nature-relation of "classic" caring-, medical healing- and breeding-practice.[20]

---

[20] Habermas says: "Wiederum andere Einstellungen erfordern die Praxis des Bauern, der das Vieh hegt und den Acker bestellt, die Praxis des Arztes, der Krankheiten diagnostiziert, um sie zu heilen, und die Praxis des Züchters, der die vererbbaren Eigenschaften einer Population nach eigenen Zwecken ausliest und verbessert. Gemeinsam ist diesen klassischen Pflege-, Heil- und Züchtungspraktiken die Achtung vor der Eigendynamik einer sich selbst regulierenden Natur. An ihr müssen sich die kultivierenden, therapeutischen oder selegierenden Eingriffe orientieren, wenn sie nicht fehlschlagen sollen." (Habermas 2001: 81).

It is this equilibrium-oriented concept of "normal" and "natural" practice, originating in Aristotle's difference between "natural" and "artificial", which seems to be endangered by the "logic of intervention".[21]

This process of levelling out the differentiation between "naturally" and "artificially generated" can be regarded as morally relevant, if it in fact endangered human nature itself. All those types of action, which show a specific reduction to pure interventionalist technicalisation are to be avoided and the identity of the human genus (biologically correct would be "species" at this point – "genus" is not a "natural" but an "artificial" category) could be guaranteed.[22]

However, this argument fails, if our knowledge on the constitution of organisms[23] can be shown to necessarily depend on this type of interventionalist causality – as it in fact does.[24]

---

[21] "Je rücksichtsloser nun die Intervention durch die Zusammensetzung des *menschlichen* Genoms hindurchgreift, umso mehr gleicht sich der klinische Stil des Umgangs an den biotechnischen Stil des Eingriffs an und verwirrt die intuitive Unterscheidung zwischen Gewachsenem und Gemachtem, Subjektivem und Objektivem – bis hinein in den Selbstbezug der Person zu ihrer leiblichen Existenz. Den Fluchtpunkt dieser Entwicklung charakterisiert Jonas so: »Als technisch beherrschte schließt die Natur jetzt den Menschen wieder ein, der sich (bisher) in der Technik als Herr ihr gegenübergestellt hatte.« Mit den humangenetischen Eingriffen schlägt Naturbeherrschung in einen Akt der Selbstbemächtigung um, der unser gattungsethisches Selbstverständnis verändert – und notwendige Bedingungen für autonome Lebensführung und ein universalistisches Verständnis von Moral berühren *könnte*." (Habermas 2001: 85).

[22] "Normative Schranken im Umgang mit Embryonen ergeben sich aus der Sicht einer moralischen Gemeinschaft von Personen, die die Schrittmacher einer Selbstinstrumentalisierung der Gattung abwehrt, um – sagen wir: in der gattungsethisch erweiterten Sorge um sich selbst – ihre kommunikativ strukturierte Lebensform intakt zu halten." (Habermas 2001: 122). The "self" is to be thought as a self mediated *by* and *through* the genus (*Homo*). "Self" then has to function within two different language games, a biological (where it does not allow normative inferences of any ethically relevant kind) and a normative one (where it does so). By assuming a necessary connection between the fundamental "biological" and the derivative "communicative" use of the term "self", we face the same inconclusive conceptual tension as was shown above.

[23] Neither Habermas nor Plessner draw a difference between organisms and living entities. Following our line of argument here, Habermas should have referred to organisms, as Plessner's concept of boundary-constituting entities explicitly applies the theory of organism, which comes from Uexküll (for further reading s. Gutmann

2. The second line of argument is based upon an application of the principle of responsibility, which our citation above already indicated (for further reading s. Jonas 2009). This application of Jonas' principle has to face two fundamental counterarguments, as it cannot be justified on grounds of an argumentative procedure. It actually rests upon a very strong metaphysical assumption on human nature and the inference of a "duty to proliferation" of human life. The second counterargument is of an immanent nature, as it emphasizes the property of procedurality, which Habermas assumes to be a necessary – and thus indispensable – condition of any type of reasoning (s. Habermas 1992). Consequently, Habermas could either stick to the ethics of responsibility by abandoning some essentials of democratic societies or he could stick to his procedural ethical framework by abandoning the (in its very core) authoritarian and paternalistic concept of responsibility.
3. Finally, Habermas presents a Kantian argument, which is based on some strong assumptions of discoursivity.

The argumentative lines 1 and 2 have to face severe counterarguments as noted; however, the third approach shows some potentials, which are tightly connected with our leading problem of artificial personality. Following the principles of discourse liberal eugenics can be characterised as unjustified:

Reciprocity and symmetry are among the core elements of justified autonomous human actions. Those actions are considered to be unjustifiable, if the consequences and results of a norm of action cannot be mutually accepted by all those persons that are potentially or actually affected. In our case, negative eugenics might be acceptable because it can be supposed that the target individual of the intervention might and would have accepted the intervention of his own account because diseases could be avoided this way. However, positive eugenics in contrast is of a completely different nature, because in this case the ultimate aim of the intervention is the determination of the phenotype according to the wishes of the parents. Even though the outcome of the assumed intervention might be

---

2004b). In consequence of this negligence, both authors fail in identifying the criteria of adequacy of their descriptions.

[24] S. Tetens 1987; Janich 1997; Janich & Weingarten 1999.

considered to be positive by most people asked, it is structurally unilateral, because the central rules of reciprocity and mutuality are violated:

1. *Symmetry* cannot be assumed, as the outcome of the intervention is irreversible (the parent's decision is irrevocably) and
2. *Reciprocity* seems to be excluded because the acceptance of the consequences by the acting party (the parents) alone does not determine the acceptability of the action, that is the acceptance by the target individual.

However acceptable this rejection of positive liberal eugenics seems to be, it still rests upon some strong empiric assumptions, as e.g. the "genetic intervention" is considered to be of a type, which is in one central aspect incomparable with other types of invention. The ultimate aim of liberal eugenics is the deliberate design of the potential childrens' properties and capabilities by their parents. Genetics has to be considered to be but a kind of means in order to realize the specified ends. This does neither imply that there do not exist other means, which allow us to realize those ends, nor does it imply that reliability of the application of genetic means is of a uniquely high level.[25] More interesting for our argument is the fact that even if genetic means were of the demanded kind, it does not (materially) follow, that other means do not show the same inferential structure. *In concreto* it seems to be dubitable at least to assume that the determination of human action by non-genetic influences (e.g. by custom or other means of social structures) is of a less thorough kind. However, even the argument as well as the counterargument is then – by definition – of an empiric nature. Consequently, we would have gained so far only the same *ceteris paribus* consequence that Sturma's argument against artificial persons already provided, by tightly connecting our ethical counterargument to the state of the art of

---

[25] Janich & Weingarten 1999 show that genetic interventions are based by a specific metaphysics which is derived from an "engineering model": organisms are considered to be the direct outcome of genetic expression; and it is this type of "genetic determinisms" which grants the expectations, connected with the application of genetic means. Those metaphysical statements neglects not only competing research programs (e.g. in terms of epigenetics) but also severe objections against monocausalistic explanations of organismic constitution (for further reading s. Gutmann & Bonik 1981).

sciences (in this case it is genetics and not artificial intelligence). However, Habermas provides a second line of argument, which is not reducible to the demand of symmetry and reciprocity in terms of discoursivity alone, but presupposed the reconstruction of persons in terms of autonomous action we gave above. This argumentation refers to a well-known version of the *categorical imperative*:

> "Denn vernünftige Wesen stehen alle unter dem Gesetz, daß jedes derselben sich selbst und alle andere niemals bloß als Mittel, sondern jederzeit zugleich als Zweck an sich selbst behandeln *solle*." (Kant 1984: 66)

From this imperative we can gain a semantic basis for developing an immanent *contradictio in adjecto*, by stating the possibility of an artificial moral agent.

4. Conclusion: Semantic Problems of Artificial Morality

Referring to the presented version of the *categorical imperative*, a person is not only defined as a rationally acting agent – this determination of a person acting as if he were a member of a rational community stays valid. The presented version is stronger insofar as it forbids the reduction of a counterpart to a pure means of the actor's actions (and determines at the same time the expectation of the reciprocal relation from counterpart *alter* to *ego*). The acceptance and acknowledgment of the *alter* as an end-in-itself determines *alter* as an end-defining and -setting agent, i.e. as an autonomous agent. In the perspective of this agent, *ego* becomes mirrored[26] as having the same expectance to *alter* as *ego* etc. The consequence of a violation of this version of Kant's imperative, stating the irreducibility of a person to a pure tool and neglecting its status as an end in itself (*"Selbstzweck"*), is then not only the irreversibility of the design-decision. This would

---

[26] The term "mirror" serves as a metaphoric expression for a conceptual development of acknowledgement, which cannot be provided here. In fact, if we assume "autonomy" not to be a property of an individual but of a person (in the sense of our argument), then personality is by definition depending on the development of the interindividual relations. Consequently, "autonomy" cannot be determined in a material way without referring to the respective state of relations, which define personality. The conceptual reconstruction of these relations would have to be the task of philosophical reflection (for further reading s. Gutmann 2004a).

be an (empirically dubitable) unique property of genetic interventions. But the main point here is the reduction of an *alter* to a pure means of *ego*'s ends. This asymmetry is indeed irreversible, and it is this irreversibility of means and ends, which lead us back to the problem of artificial personality. The case of artificial agents can be considered a mirror-problem to individual eugenics, and it leads to a destructive dilemma by stating the possibility of an artificial morally acting person. This dilemma comes into existence, because we are dealing from the very beginning – by definition – with a tool (namely the agent) that serves specific aims:

1. On the one hand the status of an agent is defined in terms of *as if*. In this case, its personality is only in "as if" form either. The agent then remains non-personal (in a strict sense) – reflecting its being designed as a tool to given and justifiable ends. The justification can be assumed to be done in terms of discourse, which represents the function of a person as a rational being. The artificial system (*AS*) then can function as a device for discursive purposes – theoretical as well as practical – and thus can provide lines of arguments, that are by definition morally. However, even in the case of moral argumentation, the *AS* remains what it is, namely "as if" person, and consequently, the moral argumentation remains in the status of "as if" morality unless a person *sensu stricto* accepts the argumentation as its own and accepts what defines autonomous decisions, namely the responsibility for the results and effects of the decision itself. The decision then is assumed to be made by the person itself – be it with or without the use of an *AS*. In the case of the use of an *AS*, even this use is the result of the person's decision and consequently the person itself[27] is to be held responsible.
2. The alternative to the first option is to state that an *as* is literally a person. In this case it is not only the capability of acting within the "space of reason" – in terms of theoretical as well as practical

---

[27] The result of this reconstruction is of course most inconvenient: there is no escaping from the final responsibility even if autonomy of decision in the strong sense is attributed completely to autonomous artificial systems; because what ever it is assumed to be, it remains the decision of the person who attributes his autonomy to such a system – and thus, this decision has to be answered for in terms of personal responsibility.

reasoning, which has to be considered to define the status of the person. Additionally, we have to take into consideration the symmetry of being an end in itself. And at this point, the immanent contradiction of an acting artificial *person* becomes lucid: the *AS* was created only as a tool to specified ends. These ends have to be justified in terms of a practical discourse as shown above. However, the "being a mean to given ends" is a status, that cannot be undone by any decision of the *AS* – if it remains what it is to be designed for, namely a tool. Even the acceptance of the *AS* as not being a tool anymore remains a decision of the person, that refers to *AS* as a person. Consequently, the author of this decision, that is the person who does not use the *AS* as a tool anymore but as an end in itself, is responsible for the decision.

In both cases, the *AS* remains what it is, namely a tool, designed according to specified ends, which have to be justified by the designer and the user of the *AS*. If our reconstruction is adequate, there are serious semantic objections concerning the hope of roboethics in order to present a new type of systems. Even if described as *moral* actor, they are only acting in terms of as-if-relations. Their morality then is exclusively a matter of metaphorical or – if explicated – of modelled normative ascription. *AS* could act morally, e.g. by taking responsibility of their rule-governed functioning if and only if we understood the term "moral" as the implementation of formalised routines and programs. This seems to be excluded semantically as long as we insist on autonomy as a defining feature of personality. Thus, there can be moral decisions of persons or there can be *AS functioning* as persons, making moral decisions – but artificial persons making moral decisions renders a *contradictio in adjecto*. If this conclusion is rejected, the retorsion can be pushed one step further, which finally clarifies the necessary reference to autonomy: the fundamental difference between "constructing an artificial person" and "educating a human being" became levelled out with the disastrous consequence that educating human beings is in fact nothing else but implementing technical and operational routines in artefacts – a conclusion which simply excludes reflected human self-understanding.

References

Anscombe, G. E. M. (1957): *Intention*, Cambridge, Massachusatts Harvard University Press.

Asaro, P. M. (2006): "What should we want from a Robot Ethic?", IRIE, Vol. 6, 12/2006, pp. 9–16.

Decker, M.; Dillmann, R.; Dreier, T.; Fischer, M.; Gutmann, M.; Ott, I. & Spiecker genannt Döhmann, I. (2011): "Service robotics: do you know your new companion? Framing an interdisciplinary technology assessment", *Poiesis Praxis* 8, pp. 25–44.

Dewey, J. (1925): *Experience and Nature*, Dover Publ., New York.

Gethmann, C. F. (1979): *Protologik*, Suhrkamp, Frankfurt.

Gutmann, M. (2002): "Aspects of Crustacean Evolution. The Relevance of Morphology for Evolutionary Reconstruction", in: Gudo, M.; Gutmann, M. & Scholz, J. (eds.): *Concepts of Functional, Engineering and Constructional Morphology: Biomechanical Approaches on Fossil and Recent Organisms*" Senckenbergiana letheia, 82 (1), pp. 237–266.

Gutmann, M. (2004a): *Erfahren von Erfahrungen. Dialektische Studien zur Grundlegung einer philosophischen Anthropologie*, 2 Bd., transcript, Bielefeld.

Gutmann, M. (2004b): "Uexküll and contemporary biology: Some methodological reconsiderations", *Sign Systems Studies* 32 (1/2), pp. 169–186.

Gutmann, M. (2010): "Autonome Systeme und der Mensch: Zum Problem der medialen Selbstkonstitution", in: Selke, S. & Dittler, U. (eds.): *Postmediale Wirklichkeiten aus interdisziplinärer Perspektive*, Heise, Hannover, pp. 127–148.

Gutmann, M. (2011): "Life and Human Life", in: Korsch, D. & Griffioen, A. I. (eds.), *Interpreting Religion*, Mohr Siebeck, Tübingen, pp. 163–185.

Gutmann, M. & Weingarten, M. (2001): "Die Bedeutung von Metaphern für die biologische Theoriebildung", *DZPh*, 49 (4), pp. 549–566.

Gutmann, M.; Rathgeber, B. & Syed, T. (2011): "Organic Computing: Metaphor or Model?", in: Müller-Schloer, C.; Schmeck, H. & Ungerer, T. (eds.): *Organic Computing – A Paradigm Shift for Complex Systems*, Birkhäuser, Basel, pp. 111–125.

Gutmann, W. F. & Bonik, K. (1981): *Kritische Evolutionstheorie*, Gerstenberg, Hildesheim.

Habermas, J. (1992): *Moralbewußtsein und kommunikatives Handeln*, Suhrkamp, Frankfurt.

Habermas, J. (2001): *Die Zukunft der menschlichen Natur*, Suhrkamp, Frankfurt.

Hacking, I. (1996): *Einführung in die Philosophie der Naturwissenschaften*, Reclam, Stuttgart.

Hartmann, D. (1993): *Naturwissenschaftliche Theorien*, BI-Wissenschaftsverlag, Mannheim, Wien, Zürich.

Hegel, G. W. F. (1986): *Phänomenologie des Geistes*, Suhrkamp, Frankfurt.

Heidegger, M. (1962): *Being and Time*, (trans. by Macquarrie, J. & Robinson), Harper Perennial, New York.

Heidegger, M. (1993): *Sein und Zeit*, Max Niemeyer, Tübingen.

Hempel, C. G. (1977): *Aspekte wissenschaftlicher Erklärung*, Walter de Gruyter, Berlin, New York.

Janich, P. (1996): "Das Experiment in der Psychologie", in: Janich, P.: *Konstruktivismus und Naturerkenntnis*, Suhrkamp, Frankfurt, pp. 275–289.

Janich, P. (1997): *Kleine Philosophie der Naturwissenschaften*, Beck, München.

Janich, P. & Weingarten, M. (1999): *Wissenschaftstheorie der Biologie*, UTB, München.

Janich, P. (2001): *Logisch-pragmatische Propädeutik*, Velbrück, Weilerswist.

Jonas, H. (2009): *Das Prinzip Verantwortung*, Suhrkamp, Frankfurt.

Kant, I. (1984): "Grundlegung der Metphysik der Sitten", in: Weischedel, W. (ed.): *Werkausgabe* Bd. VII., Suhrkamp, Frankfurt.

Lorenzen, P. (1987): *Lehrbuch der konstruktiven Wissenschaftstheorie*, BI, Mannheim.

Mill, J. S. (1974): "Utilitarianism", in: Warnock, M. (ed.): *John S. Mill, with selected writings of J. Bentham and J. Austin*, Meridian, New York.

Plessner, H. (1975): *Die Stufen des Organischen und der Mensch* (1928), De Gruyter, Berlin.

Sellars, W. (1963): "Philosophy and the Scientific Image of Man", in: *Science, Perception and Reality*, Rigeview, Atascadero, pp. 1–40.

Sturma, D. (2003): "Autonomie. Über Personen, Künstliche Intelligenz und Robotik", in: Christaller, T. & Wehner, J. (eds.): *Autonome Maschinen*, Westdeutscher Verlag, Wiesbaden, pp. 38–55.

Tetens, H. (1987): *Experimentelle Erfahrung*, Meiner, Hamburg.

Tugendhat, E. (1995): *Vorlesungen über Ethik*, Suhrkamp, Frankfurt.

Wright, v. G. H. (1971): *Erklären und Verstehen* (1991), Hain, Frankfurt.

Hermeneutics and Anthropology/Hermeneutik und Anthropologie
hrsg. von Prof. Dr. Andrea Marlen Esser (Universität Marburg), Prof. Dr. Armin Grunwald (Karslruhe Institute of Technology – KIT) und Prof. Dr. Dr. Mathias Gutmann (Karslruhe Institute of Technology – KIT)

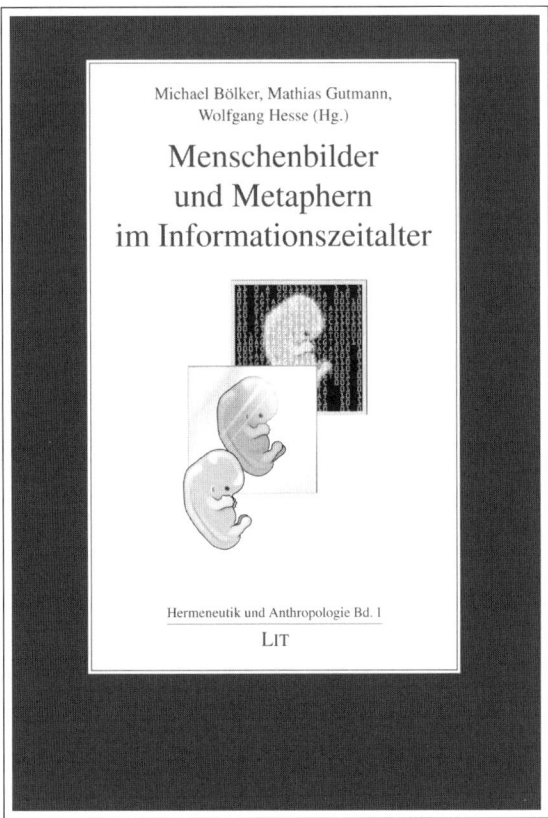

Michael Bölker; Mathias Gutmann; Wolfgang Hesse (Hg.)
Menschenbilder und Metaphern im Informationszeitalter
Information ist einer der Schlüsselbegriffe der Gegenwart. Er findet sich in zahlreichen wissenschaftlichen Kontexten und lässt sich kaum auf einen homogenen Bedeutungsbereich eingrenzen. Vielmehr spielen metaphorische Verwendungsweisen eine große Rolle in den Lebenswissenschaften und üben nicht zuletzt einen Einfluss auf moderne Menschenbilder aus. Das Ziel dieser interdisziplinären Studie ist die Analyse von informationswissenschaftlichen Metaphern und den mit ihnen verbundenen theoretischen Konzepten. Methodische Rekonstruktionen erlauben einen angemessenen Umgang mit solchen uneigentlichen Redeformen und zeigen Konsequenzen für die Bildung und Transformation gegenwärtiger Menschenbilder auf.
Bd. 1, 2010, 288 S., 29,90 €, br., ISBN 978-3-643-10310-9

LIT Verlag Berlin – Münster – Wien – Zürich – London
Auslieferung Deutschland / Österreich / Schweiz: siehe Impressumsseite

René Thun
Freiheit und Methode
Neurophilosophie bestimmt den Menschen über dessen Gehirn. Ihr methodischer Ausgangspunkt sind die Neurowissenschaften. Da die empirisch verfahrenden Neurowissenschaften Wahrnehmen und Handeln mit kausalen Mechanismen im Gehirn erklären, ist der Mensch nicht frei. Fraglich ist dabei, ob der bloße „Blick in den Kopf" eine geeignete Methode ist. Die Studie expliziert einen hermeneutischen Freiheitsbegriff, der die Gegenständlichkeit und Zweckmäßigkeit der Rede von „Freiheit" klärt. Der so gewonnene Freiheitsbegriff wird exemplarisch am Leitfaden musikalischer Praxis erprobt.
Bd. 2, 2008, 232 S., 24,90 €, br., ISBN 978-3-8258-0349-0

Benjamin Rathgeber
Modellbildung in den Kognitionswissenschaften
Kognitionswissenschaft und Kognitionstechnik gelten – nach den Worten von F. J. Varela – als die bedeutendste theoretische und technische Revolution seit der Atomphysik. Trotz dieser Relevanz scheint der Ausdruck „Kognition" selber bis heute weder eindeutig noch einheitlich eingeführt. Die hier vorgelegte Schrift versteht sich demzufolge als Beitrag zu einer Konstruktiven Wissenschaftstheorie, die auf explizite und eindeutige Weise die zentralen kognitiven Ausdrücke der Kognitionswissenschaft einführt. Dies wird im Rahmen einer allgemeinen und speziellen Modelltheorie vorgeführt.
Bd. 4, 2011, 304 S., 29,90 €, br., ISBN 978-3-643-10890-6

LIT Verlag Berlin – Münster – Wien – Zürich – London
Auslieferung Deutschland / Österreich / Schweiz: siehe Impressumsseite